A FIELD GUIDE TO

WILDLIFE

IN TEXAS AND THE SOUTHWEST

ERRATA

The butterfly on Plate 16 is an Eresimus butterfly, *Danaus eresimus*, not a Monarch butterfly.

Other books by George Miller

Texas Photo Safaris

Texas Parks and Campgrounds: North, East, and Coastal Texas

Texas Parks and Campgrounds: Central, South, and West Texas

A FIELD GUIDE TO

WILDLIFE

IN TEXAS AND THE SOUTHWEST

BY GEORGE OXFORD MILLER

TexasMonthlyPress

Texas Monthly Press, Inc.
P.O. Box 1569
Austin, Texas 78767

A B C D E F G H

Library of Congress Cataloging-in-Publication Data

Miller, George Oxford. 1943-
A field guide to Texas wildlife / by George Oxford Miller.
p. cm.—(Texas Monthly field guides)
Bibliography: p.
Includes index.
ISBN 0-87719-126-3 : $21.95.
ISBN 0-87719-072-0 (pbk.) : $14.95
1. Zoology—Texas. I. Title. II. Series.
QL207.M55 1988
591.9764—dc19 88-20054
CIP

Printed in Singapore by Tien Wah Press (PTE.) LTD

In Memory of
My Loving Mother and Father

If a man inflict a thousand ills upon a beast, it can neither ward him off with speech nor hale him into court. Therefore it is essential that ye show forth the utmost consideration to the animal, and that ye be even kinder to it than to your fellow man.

Train your children from their earliest days to be infinitely tender and loving to animals. If an animal be sick, let the children try to heal it, if it be hungry, let them feed it, if thirsty, let them quench its thirst, if weary, let them see that it rests.

—'Abdu'l-Bahá

The passage from Bahá'í scriptures is reprinted from *Selections from the Writings of 'Abdu'l-Bahá* (©1978 by the Universal House of Justice) with permission of the publisher, Bahá'í World Center Publications, Haifa, Israel.

ACKNOWLEDGMENTS

One of my major goals in writing this book was to present the latest scientifically accurate information available about the many animals I chose to include. A number of individuals who have done considerable research and devoted much of their lives to studying the various animals reviewed portions of the manuscript. I thank Alan Tennant, the author of *The Snakes of Texas*, for reading the snake sections; Dr. Merlin Tuttle, the president of Bat Conservation International, for reviewing the section on bats, and Frank Johnson and Tom Stehn of the Aransas National Wildlife Refuge for their updated information on the whooping crane. Special thanks go to Dr. C. J. Durden of the Texas Memorial Museum for his information on butterflies, Bruce Thompson of the Texas Parks and Wildlife Department for information on nongame mammals in Texas, and Ann Sorensen, formerly with the Texas Department of Agriculture Fire Ant Project, for information on imported fire ants. Many others answered questions about particular animals. Bill Lamar and Jim Stout of the Tyler Zoo helped with some of the snake photos, as did George Regmunel of Armand Bayou Nature Center and Alan Tennant; Edgar Black of Black's Barbecue in Lockhart, Texas, graciously allowed me to photograph his jackalopes; and Margaret Campbell assisted in finding photogenic ticks. I am especially grateful to my wife, Delena Tull, for the time she spent reading the manuscript and for her valuable editorial suggestions.

CONTENTS

I've Got a Question

(Everything You Wanted To Know About Animals But Were Afraid to Ask)

Once, while working at the Austin Nature Center, I received a frantic phone call. "There's a bat under the eaves of my house! Could it be one of the flesh-eating kind like I saw in the movies?" Another alarmed caller, who must have seen Alfred Hitchcock's film *The Birds,* wanted to know if the flocks of grackles in the trees would attack humans. One lady threatened to move back to Michigan if another worm-like snake invaded her bathroom. These few examples show that many people suffer from needless fear based on misinformation and ignorance.

Unfortunately animals usually suffer the most from humans' lack of knowledge. Every spring, people brought to the Nature Center jars containing snakes chopped to pieces. "Was it poisonous?" they asked. Invariably the shredded snake was a beneficial species. Harmless rat snakes, racers, and hog-nosed snakes were the most frequent victims. Once a rancher brought in three red fox cubs. "We shot the mother and didn't want the babies around," he explained. "We were afraid they would kill our cattle." The skeptical rancher would not believe that foxes eat mice, not cows. The Nature Center has a compound of cages filled with animals injured and orphaned, some accidentally, but most by people with the "chicken hawk" mentality—any bird with claws is evil, so kill it. A corollary states that any animal that has a mouth can bite, so kill it.

Not all questions we received were based on fear or prejudice. The majority came from genuinely concerned citizens trying to act responsibly toward the wild animals around them. As urban sprawl engulfs the countryside, residents find some of the animals reluctant to vacate their domain. Skunks homestead under houses, woodpeckers drum their mating call on rooftops, birds chasing their own reflections crash into windows, and the chatter of baby chimney swifts sounds alarmingly like rattlesnakes. One man in a new subdivision west of Austin reported a large furry creature gnawing on his cedar deck. To his surprise, the mysterious beast was a porcupine. An inquisitive raccoon regularly climbed down one person's chimney and scavenged through the pantry.

Various callers wanted to know if armadillos dig up and eat tulip bulbs (they eat insects), if owls suck blood from chickens (no), and if birds abandon their nests if you touch the eggs. Apparently, most parents tell their children this myth to keep the curious youngsters from harming bird nests, so the fable passes from generation to generation. In reality, most birds cannot smell, so the human scent will not keep the parents away. When I studied the hatching success of cave swallows, I measured, weighed, and numbered hundreds of eggs and young in scores of nests without causing any harm.

Humans have more sympathy for baby birds than for almost any other animal. Each spring, visitors inundate the Nature Center with "abandoned" babies. After hatching, baby birds use all their energy to reach adult size, then begin growing their main body and flight feathers. When the partially feathered juveniles cannot fit into the small nest, they take to the limbs. Their cheeps for food ring through the bushes as their parents, who are always nearby, feed and protect them. Well-meaning people mistakenly assume these babies are abandoned and "rescue" them. The Nature Center usually receives more than 100 kidnapped blue jays, grackles, mockingbirds, and other "orphans" during the peak hatching season. Feeding the screaming youngsters requires an army of dedicated volunteers.

The serious concern of the public for injured wildlife

always impressed me, and sometimes added a touch of humor to my job. One worried person wanted to know the proper first aid for a pet frog that suffered from frequent strokes, and another asked how to resuscitate a lizard. Children, whose inquisitive minds have not been programmed to accept the world as it appears, often ask the most provocative questions. "Why does a baby deer have spots?" "Why do screech owls have feathers that look like ears?" "Why does a skunk have a white stripe?" In asking these questions, children assume a purpose in the world of nature that we adults often overlook—that the features of an animal are not accidental, but satisfy specific needs.

One of the least understood concepts, yet one of the most enthralling, is natural selection. Basically stated, natural selection means that those inheritable features that help an organism survive and produce offspring will be passed on to future generations. In each generation of animals, or plants for that matter, the individuals with the physical and behavioral characteristics best adapted for survival in their particular environment will live the longest, produce the most offspring, and pass more genes to the next generation. For instance, if brown rabbits in dark woods can hide from predators better than white ones, eventually most of the rabbits in the woods will be brown.

Natural selection is one of the driving forces of evolution. The fossil record indicates that life has existed on this planet for approximately three billion years. During that time, the weather fluctuated from ice ages to tropical extremes. Riding on the earth's magma, continents floated from arctic to tropical climates, oceans periodically rose and fell, covering and exposing vast continental land masses, and countless mountain ranges arched toward the sky and eroded away to flat plains. As these changes occurred through the millennia, the plants and animals either disappeared or slowly changed in response to the new conditions. Fossils preserve the history of periods when massive extinctions occurred, such as the annihilation of the dinosaurs, as well as the gradual shift in a species' characteristics. Paleontologists, scientists who study fossils, trace the relationship of present organisms to those that long ago vanished from the earth.

The gradual shift in the characteristics of organisms as they adapt to the changing environment is called evolution. Instead of threatening our concept of the cosmos, evolution enriches it by demonstrating the creative force that governs the natural world. The chapters in this book describe how each animal has evolved to be uniquely adapted to live in its particular habitat. We can see how the size, color, food preference, hunting strategy, mating behavior, defenses, and every aspect of an organism's life are molded by the selective process of survival. The result is a world populated by life forms so diverse that they stagger the human imagination. Evolution is merely an organism's continual development to stay in alignment and harmony with an ever changing world.

The same diverseness found in the biotic world also occurs in the mineral kingdom. As a crystal grows, the laws of physics irrevocably bind every molecule in alignment with the surrounding structure. Yet, despite the rigid limitation of the physical laws, no two crystals develop identically. Crystals display a myriad array of colors, shapes, and combinations and awe us with their breathtaking beauty. Human hands have never sculpted works of art as intricate or imaginative as the crystalline structures formed by water and minerals deep within the earth. To study nature, whether animals, plants, or rocks, is to enter a realm of mystery and majesty, a world, though bound by natural laws, that remains unconstrained in the expression of diverseness.

Modern Americans, in the midst of an ongoing technological revolution, still suffer from environmental illiteracy. To many, the animals around us represent little more than names and general categories. The term "butterfly" covers the hundreds that dance on the Texas breezes, and "bird" includes the 545 species that sing in the trees across the Lone Star State. Having such a narrow awareness of the world around us is like going through life classifying people only as men and women, never knowing their names or anything about them. As we learn more about our friends, they mean more to us and a valuable and fulfilling relationship develops that enriches our lives. Our relationship with nature can be the same. The more aware we become of the animals and plants around us, the more highly we value them both for their unique characteristics and for their contribution to our enjoyment of life. As we sample the richness of creation, we begin to visualize the larger picture in which every form of life has an intrinsic place, purpose, and value.

In the world of nature, the facts are often stranger than the myths. This book explores the lives, reports the folklore, and relates the habits of some of Texas' most interesting furry, feathered, and scaly creatures. Sometimes our views of an animal mingle facts with misconceptions and obscure the truth. We hear a bird singing and conclude it is happy, when in reality it is warning other birds of its species to stay out of its territory. Anthropomorphism, attributing human feelings and motives to animals, is at the heart of most folktales and prejudice against animals. We say turtles are patient when in reality they are just slow. We consider a wolf evil for killing a deer, but praise another predator, the worm-eating robin, for being industrious in its pursuit of food. Whether we like an animal usually depends more on the human characteristics we attribute to it than on any economic impact it has on our lives. Once we learn to resist applying human values to animals, we can see the beauty in all creatures. We can view bats as one of the most marvelous and interesting mammals alive, instead of considering them ugly and despicable. Many times in this book I describe an animal with adjectives that also may apply to human characteristics, but I am not attributing human feelings, understanding, or emotions to the creature.

Wildlife

Nature is full of the unexpected and bizarre, even in urban Texas. The mammals, birds, reptiles, and invertebrates of Texas are as interesting as any in some tropical hideaway, and a lot more accessible to you and me. Read this book and get to know some of your neighbors better!

The Great Texas Wildlife Auction: Going, Going, Gone?

What did Stephen F. Austin find when he brought his first settlers to the wilderness of Texas 160 years ago? Can you imagine the High Plains so thick with bison that a traveler could ride horseback for three days with the herd stretching continually from horizon to horizon? This is just one of the images painted by early Texans. Historical records quote accounts of prairie chickens so abundant that they darkened the sun when they flew over. Settlers in South Texas could see thousands of deer in a single day. The coastal bays and marshes teemed with a seemingly unlimited supply of ducks, geese, and shorebirds. Bighorn sheep, elk, grizzly bears, wolves, and pronghorn antelope lived in abundance in the mountains and deserts of the Trans-Pecos. Prairie dog towns, some as large as 25 miles in diameter, ranged across the western third of the state.

In addition to the oceans of bison, deer, and pronghorns, the vast grasslands stretching from South Texas to the Panhandle supported millions of longhorns and mustangs, compliments of the early Spanish explorers. After the Civil War, ranchers exploited the abundant wild cattle, brought millions of dollars into the bankrupted state, and forged forever the image of the American cowboy. While cattle became king and displaced wildlife from the Texas grasslands, farmers busted the sod of virgin prairies, and the lumber industry began chewing away at the towering forests in East Texas. The rich woodlands of the Big Thicket once supported the greatest diversity of wildlife in North America. Bears, cougars, bobcats, beavers, mink, otters, wolves, Carolina parakeets, passenger pigeons, and ivory-billed woodpeckers witnessed the arrival of early Texans.

John James Audubon reported that one flock of passenger pigeons passed overhead for three hours and contained more than one billion birds. Modern-day Americans cannot imagine the abundance of wildlife that this country originally supported.

With so much wildlife, early Texans felt no need for moderation in hunting. One hunter killed 1500 deer in one year, and commercial hunters killed millions of waterfowl. A trading post near Waco shipped 75,000 deer hides to New York between 1844 and 1853. In an effort to eliminate the major food source of the Indians, hunters, encouraged by the government, exterminated the vast herds of bison. While ranchers in the west decimated bears, wolves, and other predators, farmers in the east battled the ducks, geese, and Carolina parakeets, which feasted on grain and fruit crops. A few early writers expressed remorse over the drastic decline of wildlife, but the state was slow to pass game laws. Not until 1907 did Texas appoint a game department, and it did not begin issuing hunting licenses until 1909.

The story of the taming of Texas, and the entire continent, reflects the irreparable effect of humans on an unspoiled wilderness. With the explosive growth in population, small towns became large cities at the necessary sacrifice of our native grasslands, prairies, forests, and wildlife. The consumption of the environment continues today, unfortunately with all too little concern about preserving natural areas. Wildlife in South Texas loses about 17,000 acres annually as ranchers convert brushland to pastures. Only 1 percent of the Lower Rio Grande Valley has escaped cultivation.

Lumber companies in East Texas decimate hundreds of thousands of acres of prime wildlife habitat each year. In the last decade, clear cutting of the rich forests increased by 40 to 60 percent. As the 1980s began, lumbering, pine plantations, agriculture, reservoirs, and population growth already had destroyed 63 percent of the prime wildlife habitat in East Texas. But the greatest threat to wildlife and the environment is not the god of profit, nor the production and supply of necessary goods. Ultimately, it is you and I.

Rattlesnake roundups have become a tradition in West Texas, and high society demands fur coats. These are examples of how humans exploit animals for amusement. Our pioneer ancestors found a rich land, and to survive they had to use the resources close at hand. They valued the land and animals around them primarily as objects to make their lives easier. Unfortunately that pioneer ethic lives on today in our nonrecyclable consumer society. If present trends continue, the world will lose about one species of animal or plant for every hour between now and the year 2000. The same mentality that places no intrinsic value on the life forms that share this planet with us views the world, its resources, and its peoples as opportunities for personal pleasure and profit. If we are to survive on this earth, at some point in the evolution of human society we must learn to view ourselves as stewards and not as lords and masters.

Satellite photographs reveal that the insatiable appetite of urban sprawl takes the greatest bite out of wildlife habitat in Texas. The human population continues to grow at an explosive rate, and Texas has several of the fastest-growing metropolitan areas in the nation. Each year residential developments, shopping centers, and business complexes gobble up some 200,000 acres of open space in this state.

Today we must live within the constraints of the decisions of our forefathers, whether we like the situation or not, and tomorrow our children will wrestle with the problems we create. Traditional spray and flood irrigation, wasteful but more economical than water-conserving methods, and the growing demands of cities and industry use more of the state's vast supply of underground water than rain replenishes each year. The state's water supply is rapidly drying up. Some of the major springs, like Comanche Springs in Fort Stockton, have already disappeared. Comal Springs, the largest in the Southwest, and most of the others are expected to become intermittent by the turn of the century regardless of what we do now. San Marcos Springs, with a daily output of more than 100 million gallons, forms the San Marcos River, considered to be one of the unique aquatic ecosystems in North America. Two species of fish, one salamander, and a species of wild rice live in the river and nowhere else in the world. Most of the other large springs harbor their own endemic species, which will perish if the springs dry up.

We still must come to grips with our attitude toward the environment and recognize our responsibility to future generations. We no longer can see the bison, ivory-billed woodpecker, elk, mountain sheep, or black-footed ferret in the wild, but what about the bald eagle, brown pelican, peregrine falcon, ridley sea turtle, or the other 109 threatened or endangered species in the state? We are stewards of the only wild breeding flock of whooping cranes, every golden-cheeked warbler in existence, the remaining population of Atwater's prairie chickens, bat caves with the largest concentrations of mammals in the world, and a number of fishes, amphibians, reptiles, and plants that live in Texas and nowhere else on the planet.

The decisions we make today about the environment will affect the lifestyle, recreation, and profession of our children. At this late date, we must recognize the seriousness of the threats to the environment and make every effort to save as much of the remaining natural areas in the state as possible. Texas ranks 41st in the nation in the amount of state park acreage per capita. The State Parks and Wildlife Department, with the restrictions of a failing budget, now operates 102 parks, recreational areas, and historic sites totaling about 200,000 acres, with about 20 more parks planned for the future. Private organizations, like the Texas Nature Conservancy, have preserved almost as much land as the state parks system, although most is closed to the general public. The federal government operates two national parks, a national seashore, a national preserve, ten national wildlife refuges, two national recreation areas, four national forests, two national grasslands, and scores of lakes across the state. Yet, with 98 percent of the state privately owned, those parks and preserves represent too little to save the biological uniqueness of Texas. Powered by an anticipated population increase of 46 percent during the last two decades of the century, wild Texas continues to disappear at an alarming rate.

More than 20 million Texans flock to the state parks each year. A poll conducted by the Parks and Wildlife Department revealed that more than 80 percent of the park visitors come to the outdoors to relieve the mental stress and strain inherent in our modern lifestyle. But Texans gain more than mental relief from the state's wild areas. Hunters and recreationists spend more than $432 million annually. Wild areas are a profitable, but unrecognized, business, largely ignored by real estate developers, city planners, and most private landowners.

Hikers and hunters are not the only people who benefit from wildlife and wilderness preservation. Half of our modern medicines are derived from compounds first found in nature. Researchers discovered an enzyme in the saliva of vampire bats that inhibits blood clotting; someday it may be used to treat stroke victims. What makes opossums immune to rattlesnake venom and pallid bats indifferent to scorpion stings? How do hummingbirds slow down their metabolism at night? Who can guess at the number of undiscovered cures existing in the plants and animals around us? Plants rich in hydrocarbons someday may replace the nonrenewable petrochemicals so essential to our society. The majority of plants, especially in the tropics, have never been tested for their pharmaceutical or commercial possibilities, yet the rain forests, with their rich diversity of plants and animals, probably will have disappeared by the turn of the century.

So the loss of wildlife habitat affects, not just some obscure species of bird or fish, but every one of us, and our children as well. My children cannot experience the relaxation of hiking in a wilderness setting along Barton Creek near downtown Austin. Despite a scenic easement ordinance, condominiums, office buildings, and homes line the cliffs overlooking the beautiful stream. My camping spot in the Piney Woods near Beaumont is now an exclusive housing development. The story of lost wilderness is like one of lost innocence: once gone, it can never be regained. But by proper city planning and growth management, by preserving strategic natural areas, by realizing the indirect benefits of wild areas, and most of all by developing a perspective that values the plants and animals around us, we can retain some of the remnants of our environmental heritage.

Though the investment of preserving wilderness pays immediate psychological and economic dividends, the long-run ecological benefits far outweigh the short-term results. At stake is not merely a few species but the health of the entire environmental system that we depend upon for survival. When we act to preserve a species, the whole ecosystem benefits. Like an investment portfolio, diversity means stability and protection against bankruptcy, and a diversity of species in an ecosystem indicates a healthy and stable environment. The failure of vast numbers of species, as we witness today, indicates that serious problems exist in our system. We don't have to wait until lakes, rivers, and forests die from pollution and acid rain to determine that these problems threaten our own existence. We can observe the gradual disappearance of wildlife and plants and diagnose the problem. The symptoms of disaster now unmistakably cloud the atmosphere, shadow the rain forests, and wash ashore from the seven seas. As astronauts on this spaceship earth, we must recognize that our life-support system is rapidly failing. Now is the time to sound the alarm and unify our efforts to save ourselves from ourselves.

Mammals

Sperm whales dive into the ocean depths 10,000 feet beneath the surface, and Tibetan yaks climb mountain peaks 20,000 feet above sea level. The one-to-two-inch, 0.09-ounce Mediterranean pigmy shrew chases earthworms through their burrows, while the 110-foot, 26-ton blue whale plies the ocean waves. The cheetah can sprint 60 mph, yet the sloth hangs motionlessly upside down staring at the sky above its feet. At birth, ten opossum babies hardly fill a teaspoon, while a newborn elephant weighs 260 pounds. Humans, with an intelligence unmatched in the animal kingdom, have harnessed the laws of nature and modified much of the earth's surface to meet our needs. Despite the diverseness of these many animals, they all have something in common: they are mammals.

All mammals have hair and a four-chambered heart, feed their young milk, and regulate their body temperature at a constant level. Beyond these characteristics and other common anatomical features, the appearance, behavior, lifestyle, and diet of mammals differ in every imaginable way. They run, crawl, fly, swim, burrow underground, and live in trees. Some live one to two years, others in excess of a hundred years. These versatile creatures have exploited the most extreme environments existing on this planet. They inhabit the hottest deserts and the frozen arctic, the arid wastelands and the watery depths.

According to fossil evidence, mammals developed from primitive reptiles some 200 million years ago. Only a few fossilized teeth and jaw fragments remain of the earliest mammals. Another branch of reptiles, the dinosaurs, began developing about the same time and eventually became the lords of the land, air, and sea. Plant- and animal-eating dinosaurs from the size of a cat to the giant *Tyrannosaurus rex* dominated the earth for about 125 million years, while mammals remained insignificant mouse-size creatures. Finally, after spending eons on the sidelines, mammals got their chance to become the star performers.

During the late Cretaceous period, 65 million years ago, the large terrestrial, marine, and flying reptiles, as well as about 70 percent of all other animal species, vanished. The massive extinction of species has been attributed to a variety of causes, including changes in climate and vegetation, habitat destruction from mountain formation, and the cataclysmic impact of a meteor, the hypothesis currently most favored. The absence of the dinosaurs allowed mammals to begin a rapid development. A tremendous increase in mammal species occurred during the following 40 million years. Hooved animals developed, bats took to the air, whales and dolphins turned to the sea, and elephantlike animals with trunks appeared. The grazing animals diversified into a variety of sizes, and dog- and catlike animals became the major predators. The descendants of only a few of these ancestral families exist today. We know the rich mammal fauna of that time period primarily from studying the fossil remains.

Modern mammal families began to develop during the Miocene epoch, from 23 to 6 million years ago. Horses and camels originated in North America and dispersed to the rest of the world, and elephants, antelopes, and giraffes evolved in Africa. The explosive development not only of mammals but also of insects and birds is closely related to the variety and abundance of flowering plants that evolved during this period. The flowers and fruit provided a new and abundant source of food. Today 75 percent of the mammal families in the Northern Hemisphere date from the Miocene, and about 50 percent throughout the world.

Three types of mammals exist in the world today, each with radically different methods of reproduction. The most widespread are the placental mammals, whose young develop within the mother's body before birth. The babies are born live and may require extensive parental care before they become independent. The marsupial mammals have a pouch in which the young develop after an immature birth. The young crawl into the pouch and nurse until they have developed enough to emerge. Placental mammals have a relatively larger brain than marsupials, and generally displace marsupials when the two compete for resources.

About 3 million years ago, the Isthmus of Panama connected North and South America, allowing previously isolated animals to associate and compete with each other. The placental mammals from the north crossed the land bridge and displaced the predominantly marsupial mammals of the south. The opossum family is the only surviving marsupial family in North and South America. In Australia, which had no placental mammals until they were introduced in modern times, a diversity of marsupial mammals fill all the environmental niches that placental mammals occupy in the rest of the world. Australia is also home of the third class of mammals, called monotremes, which consists of two species, the duck-billed platypus and the spiny anteater. These bizarre animals lay eggs, but have primitive mammary glands.

What enabled mammals to assume such a dominant role in the world after the dinosaurs disappeared? Mammals developed, diversified, and covered the face of the earth because they could adapt, both behaviorally and physiologically, to environmental extremes and take advantage of new situations. Unlike reptiles, mammals maintain a constant body temperature independent of the climatic fluctuations. This permits longer daily and yearly periods of activity. Maintaining a constant internal temperature requires expending considerable energy. Therefore, food, more than temperature, limits the success of mammals and where they can live.

The development of hair is closely related to a regulated body temperature. Hair insulates an animal from cold, as well as performing other vital functions. It protects the skin from injury and increases the tactile sense. Rodents and cats have specialized whiskers to sense touch. In modified forms, hair provides an effective defense. Horns and claws are derived from hair, as are the protective scales on the pangolin and the shields on an armadillo.

A large brain gives mammals the intelligence required to discover and exploit new sources of energy in the environment, which has allowed them to diversify and fill every available niche. When a food source is particularly abundant and dependable, an animal can specialize, as the anteater does, and reduce its competition with other animals. Some mammals, like coyotes, take a smorgasbord approach and feast from a variety of sources, increasing the likelihood that some food will always be available.

Respiration and other bodily functions consume about 85 to 90 percent of a mammal's food energy, leaving 10 to 15 percent available for growth. An animal that grows large must have a plentiful supply of food; thus, the largest mammals dine on plants, the primary and most abundant source of energy available. A carnivore, or meat eater, lives on the fraction of food energy in the ecosystem that is stored in the bodies of its prey. With less energy available, carnivores tend to be fewer in number than the animals they eat and often prey heavily on much larger animals. Omnivores, like raccoons, bears, and humans, eat both plants and meat and are some of the most successful animals at survival because they rely on a variety of food resources.

From about 30,000 to 5000 years ago, a massive wave of extinction swept across North America and Europe, annihilating most of the large mammals. Creatures that had withstood the climatic rigors of four glaciations suddenly disappeared. The horse, camel, mastodon, woolly mammoth, giant bison, giant beaver, ground sloth, imperial elephant, dire wolf, sabertooth cat, and almost every other large animal vanished. Only a few medium-sized mammals survived on a continent once as rich in wildlife as the African savannahs. We may never know for certain what caused the demise of the large mammals, but since humans migrated into North America during this time, many scientists speculate that the animals were hunted to extinction.

Though we may speculate about the influence of humans on past extinctions, we have no doubt concerning the animals that disappeared in recent history. Between 1600 and 1900, approximately 75 species of mammals and birds vanished worldwide, about 1 every 4 years. Between 1900 and 1980, the rate averaged 1 per year, then accelerated to 1 to 3 per day during the early 1980s. If all animals and plants are considered together, the current rate of extinction has increased to approximately 1 species per hour. We may see 1 million species of plants and animals disappear during our lifetime.

Like the dinosaurs of the past, mammals are the lords of the present age. Today about 4230 mammals in 122 families inhabit the earth. Rodents comprise 42 percent of the species and bats 23 percent. We might like to think that mammals—especially one species, humans—are the most advanced animals on the earth. However, insects and other forms of life easily match the diversity and adaptability of mammals. Mammals comprise only about 10 percent of the vertebrates, or animals with backbones, and vertebrates only 4 percent of the animal species in the world. Humans may be the dominant species today, but, with all life linked together as partners sharing a common homeland, nature offers no guarantees about the future. The immediate fate of the human species, and perhaps of most life on the planet, depends on whether we can avoid a self-inflicted catastrophe surpassing the one that eradicated the last great lords of the earth.

The next 13 sections describe a selection of the 156 species of mammals that have been recorded in the state of Texas (and one fictitious creature). Most also inhabit the southern and western United States, and some range through Mexico to South America and north into Canada. Texas has 50 species of rodents, 30 bats, and about 18 marine mammals; 22 species are so rare that they have been sighted fewer than 5 times in the last 80 years. Though every animal has its unique features, and sometimes bizarre adaptations, I have chosen those that I think have the most interesting stories to tell, and whose lives have influenced and been influenced the most by humans.

The Armadillo: Texans' Newest Native Son

Nine-banded armadillo
Dasypus novemcinctus

The sign proclaimed in hand-painted letters "Armadillers for Rent." Curious people crowded in front of the pen of scampering animals at an "armadillo race" in Austin.

"Are they mammals or reptiles?" asked one puzzled onlooker.

"Neither," said the man renting the animals for the race, "they're just plain ol' dillers!"

Today people call them "dillers," but during the Great Depression folks referred to them as "Hoover's hogs." In the hard times, these unusual critters became a culinary delicacy, if not a necessity, for many people. In the past decade, Texans have rediscovered the lowly armadillo and enthusiastically adopted it as their native son. Roadside vendors and exclusive specialty stores sell armadillo hats, purses, belt buckles, and bumper stickers. Underground comic books and nationally televised beer commercials glorify the animal, and the Texas Legislature even designated the armadillo as the official state mammal. Scores of towns sponsor armadillo races, where, to the roar of intoxicated crowds, adults on hands and knees chase terrified animals across a finish line. Ironically, a society that fines and even jails people for mistreating a dog or cat is blind to the flagrant abuse of a native animal like the armadillo. The care and protection that the law affords and that society expects for our pets should extend to wildlife as well.

Why armadillos have become so popular with modern Texans puzzles even the most astute psychoanalysts. With pea-size eyes, a piglike snout, and the uncanny ability to get massacred on highways, armadillos do not exactly inspire awe or project an image about which Texans can brag. If armadillos could brag, however, they would put Texans to shame. Long before humans migrated into North America, these animals homesteaded in what eventually became the Lone Star State. They have rooted through Texas soil for more than one million years, which makes today's animals 250,000th-generation Texans. Not even the most boastful native Texan can match that genealogy!

Armadillos are as independent as a West Texas rancher, and about as sociable as a cattle rustler. They have little or no interaction with their kin or other animals. They probably never noticed the Spanish conquistadors who claimed the land north of the Rio Grande, nor the Anglos who fought both the Mexicans and Indians. Armadillos just wiggle their piglike noses in the sand and contentedly let the world go by.

An armadillo can easily afford an unconcerned attitude about the changing world. Like its South American cousins, it wears the family armor that protects it from most predators and the thorny bushes through which it scampers. The armor is actually a layer of bony plates covered with horny tissue. Its scalelike skin adds to the armadillo's bizarre appearance, and gives this strange mammal a reptilian look. The protective shield covers its shoulders and hindquarters, and nine movable bands protect its back. With such an effective armor, the armadillo does not need speed, keen vision, or super intelligence for defense. Unlike its South American relatives, however, the nine-banded armadillo cannot curl into a ball to cover its unprotected abdomen.

An armadillo responds to danger in a manner opposite to the macho image of a brave Texan. When alarmed, it stands on its hind legs, holds its nose in the air, and squints its beady eyes. If truly frightened, an armadillo jumps straight up and comes down running. This strategy works well, except that the animal often forgets the danger in a few seconds and nonchalantly resumes its rooting activities.

Perhaps Texans secretly identify with the way an armadillo copes with existence. This unconcerned creature spends its life wandering aimlessly through the woods unaware of man or beast. A contented armadillo has its nose several inches in the soil sniffing and rooting for bugs. The only well-developed sense these nearsighted creatures require is smell, necessary for finding food.

Perhaps the one thing armadillos really have in common with modern Americans is that both travel far from home. Since early naturalists first mapped their range in 1890, these rambling creatures have spread from Central Texas into New Mexico and Oklahoma, and east to the Mississippi River. Armadillos purportedly cross streams in the most straightforward manner—by walking underwater—but they needed the help of a Marine sergeant to ford America's biggest river. After World War I, a soldier stationed in Texas took two armadillos across the Mississippi River to Cocoa Beach, Florida. Armadillos, the sergeant soon discovered, make miserable pets. He released the pair, which immediately began to populate the state.

Soon the tanklike creatures had invaded Alabama and Georgia. Armadillos are exceptionally good at populating. Females always give birth to identical quadruplets. Born with their eyes open, the babies can scramble after their mother in just a few hours. Only freezing weather seems to stop the unassuming emigrants' northward march. Temperatures below 50 degrees F cause these almost hairless creatures to shiver violently.

To protect themselves against cold weather, armadillos insulate their burrows, which they dig in great numbers, with leaves and other vegetation. How does an armadillo gather leaves? Just like you or I do—it rakes them into a pile and picks them up. With its front paws full, it hops awkwardly backward to its home. Skunks, snakes, mice, and other animals often make comfortable homes in the extra armadillo burrows.

A warm burrow and a leaf-covered forest floor meet an

armadillo's needs. These creatures spend the evening, and sometimes midday, hours rooting through the leaf mulch and soft soil for insects and other small critters. They grind up beetles, scorpions, spiders, and centipedes with 32 peg-like teeth. Armadillos eat many bugs highly destructive to crops and range grasses, yet a misguided rancher near Austin bragged that he had shot more than 500 in two years. Sportsmen sometimes contend that armadillos decimate the quail population by eating eggs. A four-year study, however, showed that invertebrates comprise 93 percent of an armadillo's diet. Armadillos damaged only 9.7 percent of the quail nests studied, a number too small to affect the bird population.

After enjoying a million years of peace, prosperity, and obscurity, armadillos face trials unforeseen by nature. Today these animals must contend with being kidnapped for armadillo races, attacked by four-wheeled predators, and considered nature's biggest joke. The armadillo's four genetically identical siblings are a boon for scientific and medical research that requires identical subjects for testing. Medical researchers have discovered that armadillos carry a leprosy virus similar to the one that infects humans. Between 5 and 12 percent of the armadillos within 100 miles of the Texas coast and in South Texas carry the virus. Only about 2 percent of the rest of the state's armadillo population tests positive for leprosy. The Federal Centers for Disease Control in Atlanta warn that, though it is extremely unlikely, humans may be able to contract leprosy from armadillos. At least one case, a man who used armadillo meat to make sausages, has been reported. Nonetheless, Texans are undaunted in celebrating their newest and most popular native son and universally ignore this danger.

Bats: Mammals of the Sky

Mexican free-tailed bat,
guano bat
Tadarida brasiliensis

Texas has more bats and more species of bats than any other state in the nation. This fact would horrify the many people who rate bats along with rattlesnakes, cockroaches, and tax collectors. When an Austin newspaper columnist solicited letters on "squeamish" creatures, responses about bats showed the most hysteria. Bats probably suffer more than any other animal from an unwarranted reputation based on folktales. At the sight of the winged mammals, normally rational people react with a fear born of superstition, prejudice, and misunderstanding.

Bats are harmless, extremely beneficial, and some of the most scientifically interesting creatures on the earth, yet they probably will never receive the widespread admiration they deserve. We readily value koalas, baby seals, and pandas, which live thousands of miles away, but not the furry creatures that catch flying insects in our own backyards. The habits of bats perfectly combine the elements of myth and superstition. First of all, they flutter through the night, enough to scare the child in us all. Second, many have contorted faces, which we interpret as evil and sinister. Last but not least, some members of the clan feed on blood. If the vampire myth were not enough, the exaggerated rabies scare in recent decades has guaranteed bats a bad reputation.

At one time Texans welcomed bats as honored guests, and even built special belfries for them. In the early 1900s, mosquito-borne malaria threatened much of Texas, and health officials considered bats an effective mosquito control measure. A San Antonio health inspector built 11 towers, each 50 feet high, to attract these tiny creatures with large appetites. With each bat catching at least 5 grams or more of insects per night, the Texas population of guano bats alone consumes approximately 144,000 tons of insects annually. Guano, or bat excrement, makes a rich fertilizer, and several million bats leave tons behind when they fly south for the winter. Ranchers with bat caves shovel up the remains and pocket a tidy Christmas bonus after the bats leave. During the winter of 1985, the owners of Bracken Cave near San Antonio extracted 150 tons of guano.

During World War II, bats stimulated the imagination of the military. The Air Force devised a devilish scheme to burn Japanese cities. According to the plot, bats with incendiary bombs attached to their bodies would be refrigerated until they went into hibernation. While in the torpid state, the booby-trapped animals would be packed in special boxes and dropped over the target city in slow-falling parachutes. If the plan worked, the bats would regain consciousness during the fall and immediately seek a dark refuge under the eaves of the Japanese buildings. When the bombs went off, the Japanese and the bats would never know what hit them.

The military finally perfected the technique, but only after thousands of failed bat bombs. The tiny kamikazes burned down a model Japanese city constructed in the desert, as well as an auxiliary air station near Carlsbad, New Mexico. But the Japanese never had to endure the suicide squadrons of flying mammals, for the atomic bomb proved much more efficient at death and destruction.

The primary occupant of the numerous limestone caverns and sinkholes of Central and West Texas is the Mexican free-tailed or guano bat, *Tadarida brasiliensis*. Guano bats outnumber all other bats in the state combined. Each spring, 100 million of the ⅓-ounce flying mammals migrate to Texas from the tropics to raise their families. Bracken Cave hosts 20 million, the highest concentration of mammals on earth. The bats from this cave consume 125 tons of flying insects nightly. Nine other caves attract from 4 million to 10 million bats each, and several smaller caverns boast populations of up to a half-million guano bats. These bats adjust to suburbia as well and roost under bridges, under eaves, and in attics. The largest urban bat colony in the world lives under the Congress Avenue Bridge over Town Lake in Austin. Three-quarters of a million Mexican free-tailed bats roost there, three times as many as live in Carlsbad Caverns, emerging at night to eat insects attracted to the city lights.

Most of the guano bats living in the caves and sinkholes from February to November are females and their young. The males live in separate caves or find roosting sites in cities. The giant nursery colonies fly from cave entrances like smoke curling through the evening air. The bats fan out and forage as far as 50 or more miles away. Air Force radar tracked the bats spiraling from Bracken Cave to an altitude of 11,000 feet before the flock dispersed, riding the high-altitude winds to their feeding grounds. A helicopter once followed a flock flying 60 mph.

Large nursery colonies of Mexican free-tailed bats may double their size with millions of infants in the summer. In mid-June, the females give birth to one baby. The young huddle together to keep warm while their mothers forage for insects. Amid the millions of returning females, the babies scramble for their mothers, who unerringly find their offspring. In general, bats have fewer babies than any other mammal their size. Nesting in large congregations and having only a few babies per litter make colonial bats extremely vulnerable to losses caused by human vandalism and disturbance.

Guano bats prefer the open, semi-desert habitat characteristic of the Southwest. They live from Houston west to the Pacific coast and winter in Mexico, Central America, and the Caribbean. They forage 15 to 100 feet or more above the ground catching their primary food, moths, and mosquitoes and any other insects that are available.

Texas has about 29 bat species. Not all bats live in large

colonies. The red bat, *Lasiurus borealis*, roosts singly and protects and nurses her one to three young under a leaf on a tree limb. Red bats live east of the Rockies from southern Canada to Mexico and into Arizona and California, and winter farther south. Besides the guano and red bats, the most common include the cave bat, *Myotis velifer*; Georgia bat, *Pipistrellus subflavus*; western pipistrelle, *Pipistrellus hesperus*; pallid bat, *Antrozous pallidus*; and hoary bat, *Lasiurus cinereus*.

The only vampire bat ever found in Texas did not live to see the light of the next full moon. Wrapped in its capelike wings, the creature lived in an abandoned railroad tunnel west of Del Rio. The vampire died, not by a stake in its heart, but in a researcher's preserving jar. Today, the blood-eating bat, a hairy-legged vampire, *Diphylla ecaudata*, sits on a museum shelf. These vampire bats typically inhabit tropical and subtropical forest land and lap, not suck, the blood of birds. Though other tropical species of vampires feed on mammal blood, they rarely choose humans as hosts.

Contrary to the popular expression "blind as a bat," these creatures of the night possess excellent eyesight, though vision is not their primary means of finding prey. Bats have the ability to locate objects by listening to high-frequency, or ultrasonic, echoes. They send out high-frequency beeps and read the reflected echoes like a radar screen. Though some shrews and rodents also use ultrasound, none approach the sensitivity of bats. Bats can see with sound as well as humans see with vision. The parabolic shape of a bat's ears helps focus the echoes. Bats can detect large background objects 100 yards away, and discern the size, shape, and even texture of closer objects.

The frequency pattern used by a bat depends on the terrain. Guano bats forage in open shrubland and high above the ground, and are not concerned about flying into trees or other objects. They search for prey using a constant frequency of about ten beeps per second, which indicates the presence of an object but not much else. When pursuing prey, the bat switches to a faster, frequency-modulated (FM) signal that reveals much more information about the object. Bats foraging among trees or other obstructions alternate their search signal between constant and modulated beeps.

Before high-speed photography, people generally assumed bats always caught insects in their mouths. Now we know that they frequently snare insects with the fingertips of their webbed wings, transfer them to the cupped membrane connecting their tail and hind legs, and eat them in flight. But not all bats eat flying insects. Some tropical species specialize in small mammals, fish, calling frogs, fruit, nectar, and crawling insects.

Bats in North America avoid the scarcity of flying insects in the winter by either migrating or hibernating. Most hibernating bats seek a cave or crevice deep enough to maintain a stable air temperature above freezing throughout the winter, but some, like the red bat, hibernate in trees. A bat's ability to lower its heart beat and metabolism (how fast its body uses food) is a marvel of adaptation. The tiny creatures drop their heart rate from up to 600 beats per minute to as low as 10 per minute. At an air temperature a few degrees above freezing, bats have the lowest metabolism of any mammal. If the temperature drops below freezing, bats must burn stored fat to generate heat, and at higher temperatures their bodies naturally use more energy. Bats that remain in Texas hibernate only a few months and may break their torpid state on warm days, but those in northern latitudes hibernate as long as eight months.

Instead of hibernating, many species of bats migrate south for the winter to areas with abundant food. Migrating Texas bats enjoy the tropical climate of Mexico, Central America, and the Caribbean islands from November to February. Marked guano bats flew as far as 800 miles from their Texas nesting site into Mexico.

Even when not hibernating, a bat has the surprising ability to lower its metabolism while sleeping during the day. As it placidly hangs in a protected retreat, its body temperature approaches that of the surrounding environment, usually 65 to 70 degrees F inside a cave. Most bat-size mammals live only 1 to 2 years, but bats have survived for 30 years. Scientists once thought the ability to slow down bodily functions on a daily basis might account for the unusual longevity of many species of bats. Recent studies, however, have shown that some tropical bats that maintain a high metabolism live more than 20 years.

The 29 species of bats in the Lone Star State live in almost every imaginable habitat. They inhabit the deserts, forests, mountains, plains, and cities. They hang upside down in caves, cracks, fissures, hollow trees, buildings, under bridges, and from branches and twigs. The Seminole bat, *Lasiurus seminolus*, roosts in clumps of Spanish moss. Some forage exclusively over water, others over grass, under the forest canopy, or above the treetops. Most pursue flying insects, but the pallid bat lands on the ground and catches beetles, grasshoppers, crickets, and scorpions. Surprisingly, the sting of deadly scorpions that can kill humans does not affect the tiny pallid bat. The Mexican long-tongued bat, *Leptonycteris nivalis*, feasts on nectar and pollen in the mountains of Big Bend. The agaves growing so abundantly in the desert depend almost exclusively on bats for pollination. Every species of bat has its preferred habitat, roosting sites, foraging pattern, and diet items. Such specialization enables many different species to inhabit the same area without competing with each other.

Even though bats live by the millions in Texas, many experts consider some species of the small mammals threatened with extinction. In recent years, their numbers have decreased at an alarming rate. The decline once was attributed mainly to pesticides and habitat destruction, but disturbance from cave exploration and vandalism appear to represent even greater threats. Bats have deserted caves that once hosted enormous populations. One thoughtless person could cause a million bats to abandon a nursery cave, leading to their untimely death. If disturbed during hibernation, a bat loses 20 to 30 days of fat reserves before it can return to its deep sleep. Several disturbances during a winter could doom the entire population of bats in a cave.

Bats in Texas enjoyed a beneficent image and were protected by state law from 1925 to 1957. As the state became urbanized, attics, eaves, and buildings proved to be alluring substitutes for the disturbed caves. The greater contact with humans increased the chance of infection from a sick bat, and a paranoia worse than the vampire superstition gripped the populace. The fear of rabies brought these beneficial animals out of favor, despite the fact that many more people are injured by pet dogs and cats than by bats. In 1985, 1000 people suffered from dog bites in Austin and Travis County, and one child died. Only two people, both researchers knowingly taking unusual chances, have died from bat-borne diseases in Texas.

The first bat I saw as a child was wrapped in a cloak of wings, hanging from a twig just above my tree house. It was probably a red bat, which commonly roosts in trees across the state. How wonderful, I thought as I plucked it from its perch and ran inside to show my mother. She shrieked as though I were holding a rattlesnake, and she made me wash my hands for at least ten minutes. For days she watched me to see if I was foaming at the mouth.

Recent advances in identifying rabies have shown that each animal species carries a different strain of the virus. Tests can determine exactly which animal has transmitted the disease. Bats seldom spread the dreaded virus to other animals. The state health department reports that no rabid animals tested in Texas (as of 1986) contracted the disease from bats. In other states, only a cow and a cat have been infected by sick bats. Bats accounted for only 16 percent of the rabid wild animals tested statewide in 1984. Scientists working with bats or spending time in bat caves often take rabies immunization shots as a precaution. Probably less than one tenth of one percent of the bat population has rabies, so bats pose little threat to the public, as long as the obviously weak and unhealthy ones are avoided. If a bat can be caught, it is probably sick and should not be handled.

Because of the worldwide threat to bats, Bat Conservation International was founded to educate the public, prevent extinction and habitat loss, and to ensure survival of viable populations. Members of the organization receive a newsletter and educational materials—including backyard bat houses—and can participate in special programs and outings. The BCI address is P.O. Box 162603, Austin, Texas 78716.

The Cougar: The Reclusive Cat

Mountain lion, cougar, panther
Felis concolor

Startled passengers in passing cars stared at the four of us crammed in the cab of the pickup truck. A policeman pulled up and momentarily considered stopping us but changed his mind. Three of us were a little nervous ourselves about the fourth passenger, an adult mountain lion. Though declawed and defanged, it could have killed us all in a matter of seconds. Fortunately we all survived the trip to the television station, where the lion starred in a local talk show. By appearing on TV, we hoped to dispel some of the common misconceptions about this docile, but powerful, cat.

Mountain lions are the most widely distributed land mammal in the world, yet one of the most misunderstood. These 8-foot, 180-pound beasts strike terror in the hearts of people from the Yukon to the tip of South America. Most consider cougars blood thirsty killers that scream in the night and silently stalk human victims, yet since colonial times these cats have killed fewer than a dozen people. The only recent attack in Texas occurred in the Chisos Mountains of Big Bend National Park in August 1984. A family of hikers came upon a juvenile that had no fear of humans and appeared to have once been a pet. When one of the children screamed and began to run, the cougar attacked with the instinctive pursuit response to running prey. The lion severely injured the eight-year-old child and would have killed him if his father had not wrestled him loose. The next day, rangers tracked the cougar with dogs and killed it. Despite the proximity of lions and park visitors, a cougar had never before attacked anyone in the park, which has 18 to 20 lions in its 1100-square-mile boundary. In the spring of 1986, cougars attacked two children in California. Since mountain lions normally prey on animals the size of small children, preschoolers naturally stimulate the lions' chase and kill instincts.

Until the last two or three decades, even the National Park Service considered mountain lions worthless vermin and exterminated them. Only in 1970 did the last western state repeal its bounty laws on these cats. Today, Texas alone offers no protection to cougars; all the other western states classify them as game animals with a regulated hunting season. Ranchers hate pumas, panthers, cougars, or whatever they may be called, and kill them whenever possible.

Despite their evil reputation, mountain lions would rather flee than fight. In a lifetime of tracking, killing, and capturing 668 cougars, a famous western hunter saw only one lion that had not been treed by dogs, and he never heard the animals scream. The nocturnal shrieks usually come from owls, or bobcats in heat. Cougars detest barking dogs, and they will climb a tree to avoid a single hound, even though they could easily kill the canine. Trained dogs track the lions and hold them in a tree until the hunter arrives.

The unbelligerent, reclusive nature of these large cats contributes to the myths and half-truths repeated from generation to generation. While some Indians elevated cougars to representatives of the deity, early explorers considered them diabolic. Any animal so secretive must be hiding its evil intentions, so they reasoned as they stared into the night and told imaginative stories based on fear and conjecture. Humans justly should fear these powerful cats that can kill with a single slap—a cornered lion will fight to the death. But cougars are threatened more by humans than we are by them.

The pioneers moving west across the continent pushed the mountain lions before them, replacing the wilderness with farms, ranches, and cities. Hunting and loss of habitat eliminated the lions from most of their historical range, which originally included every state. In North America, the large cats survive primarily in the deserts and mountains of the West; the only breeding population east of the Mississippi lives in Florida.

Once common from shore to shore, these adaptable felines lived in mountains, forests, prairies, plains, swamps, and deserts. Their range corresponded to the distribution of their primary food—white-tailed deer in the East and mule deer in the West. As predators, cougars played an important role in keeping the deer population in balance with food resources. Decades of study have shown that the lions invariably kill the sick, feeble, and surplus young deer in the herd, leaving the healthy and strong animals to breed and care for the young.

The beneficial relationship between prey and predator was illustrated on the Kiabab Plateau on the north rim of the Grand Canyon. In an effort to "manage" the deer herd, the government systematically killed all cougars and wolves. With no natural controls, the deer population rose from 4000 in 1907 to 100,000 17 years later. Finally, with the forest abused and overgrazed, the starving deer died in great numbers until less than the original number survived.

An adult mountain lion kills between 35 and 100 deer annually, usually closer to the smaller number; these opportunistic hunters also will eat anything available. They savor porcupines and kill the quilled animals whenever they find them. Cougars eat mice, raccoons, and other small mammals, as well as insects and grass. In the deserts, they regularly hunt javelinas. Hungry cougars hunt by searching and stalking and by waiting and pouncing. They prefer country with heavy cover to conceal their movements and precipitous terrain with rocky ledges and escarpments so they can lie in ambush.

When a deer gets within range, the cat leaps toward the prey and pulls it down by grasping its shoulders and biting its neck. Cougars can bound with lightning speed for several jumps, but they do not have the endurance to win a race and seldom chase prey. The prey often dies instantly with a broken neck from the lion's powerful assault. The lion then drags the animal to a secluded spot and eats its meal. When full, the cat covers the carcass with leaves. It may return the next day to feed again, but mountain lions will not eat rotten meat. Coyotes, bobcats, eagles, vultures, and other less picky eaters dine on the leftovers.

A hungry cougar, like any opportunistic hunter, eats the food most available and easiest to catch, whether mice and deer or sheep or goats—thus, the eternal conflict between stock raisers and large predators. Ranchers also despise mountain lions because the cats occasionally kill much more than they can eat. The multiple attacks result not from a sadistic love of killing but from an instinctive reaction. A cat pounces on and kills prey, not in response to hunger but in response to the fleeing action of the prey. Wild animals naturally run and scatter to safety, but a herd of sheep, bred for high productivity of meat and wool, may mill around in confusion. As long as the stimulus is present, the lion will kill. If you encounter a cougar in the wild, do not run like the child in Big Bend, but slowly retreat.

Cougars are highly specialized hunters, superbly equipped to kill large animals. They have stereoscopic vision, necessary to judge distance when stalking and attacking, and their large eyes are adapted for night vision. Their canines can slash deep wounds, and their jaw muscles can crush bones. Their retractable claws can disembowel a deer or elk with one slash, and their powerful leg muscles break the prey's neck with one blow. Yet despite their deadly weapons, most cougars pose less of a threat to humans than a bull that charges any person entering its pasture.

Mountain lions prefer a life of solitude and associate only when mating. The male stays with the female during the two to three weeks that she is in heat and drives away any intruding males. Sometimes vicious fights occur, with the dominant male winning the right to father the cubs. The female chooses a shallow den under a ledge or log to have her one to six kittens. The black spotted cubs open their eyes in about ten days and nurse for two to three months. After weaning, they accompany their mother on hunting trips, and in the course of the next two years learn the hunting skills necessary to survive. During their second winter, the young leave their mother, who is ready to breed again, and strike out on their own. The juveniles may establish a home range from six to ten miles away or wander for a hundred miles until they mature and establish their own territory. The transient juveniles replace lions that have died or been killed.

Cougars mark their home range, which differs from a true territory because home ranges often partially overlap. Within that area, the cat maintains a small resting range for its exclusive use and a larger hunting range that it may share with its neighbor. The cat scrapes together a pile of leaves along a well-traveled trail and marks it with urine. Any wandering lion respects the No Trespassing sign and turns away to avoid conflict.

One morning I woke up in my desert campsite in Big Bend National Park and discovered mountain lion tracks within a few yards of my tent. Later that day, I found the lion's fresh scrape on a ridge overlooking the camp and Terlingua Creek. The cat had scraped a three-foot circle with a pile of dirt and an old rag in the center. I would have loved seeing the lion silhouetted in the moonlight along the ridge.

Wildlife researchers know little about the status of the mountain lion in Texas. Breeding populations exist in Big Bend and Guadalupe Mountains national parks, and ranchers regularly encounter the animals in the Trans-Pecos and south of the Nueces River. Scattered sightings occur in the Rolling Plains and Cross Timbers areas in North Texas and on much of the Edwards Plateau. In the early 1980s, one was seen repeatedly near Lake Travis. Most of the sightings represent transients, some probably driven from Mexico by increasing hunting pressure. The depressed economics of sheep and goat ranching have favored the mountain lion in the last decade. With fewer livestock on the range, the pressure to control the predators has decreased. Though the number of Texas cougars is undocumented because of lack of research, more of these magnificent cats may exist now in the state than in recent history.

The Coyote: The Midnight Minstrel

Coyote
Canis latrans

A startled jackrabbit raced from the bushes as we bounced down the dusty road to our camp at Terlingua Abaja in Big Bend National Park. Suddenly, from out of nowhere a coyote appeared, and the chase began. The rabbit shifted into high gear and crested a low hill with the coyote a breath behind. A few seconds later, the panting canine trotted off empty-handed, like its cartoon counterpart, Wile E. Coyote. The coyote probably knew it didn't have a chance, but what could be better sport than chasing a rabbit?

Coyotes seldom overlook anything, especially if it relates to food or danger. Their keen vision, hearing, and smell keep them tuned in to their surroundings. At the Mule Ear Peaks overlook, one came trotting over a rocky bluff a hundred yards from my car. It instantly stopped to inspect the human invaders in its desert domain. A cautious nature has saved more than one coyote from human ill will. We stared at each other for a while, I with my binoculars, he with his sharp yellow eyes. As soon as I took out my telephoto camera lens, he melted into the underbrush. A minute later the coyote was a quarter mile away, resuming his daily rounds.

A coyote loping across the open desert seems a natural part of the rugged surroundings. Adults make long migrations, and the young often disperse as much as 100 miles from their parents' territory. These lean and lanky animals can jog at a comfortable gait of 10 to 12 mph for hours without tiring. They can run 25 mph with bursts of speed up to 40 mph. Coyotes will lead hunting dogs on an all-day chase if given the chance. A pack of hounds once pursued a coyote 50 miles before they dropped from exhaustion.

I found a coyote den in a sandy river bank near San Vicente, a Mexican village just across the Rio Grande from Big Bend National Park. Coyotes can dig like miners, but would rather use an old badger hole, rock crevice, or other protected place to rear their young. No one was home, but I could imagine the four to seven frisky puppies rolling in the sand, attacking each other, and tugging on mom's and dad's ears. The parents bring food to the young when they start eating solid food, after about three weeks. By nine months, the juveniles reach adult size and soon after leave the family.

Coyotes are easily satisfied. When they are tired, any shady or protected spot within their home range serves as a bedroom. A pair maintains a home range from one to ten square miles, and in some instances aggressively protects this domain. Yet these social animals often greet visitors with the same friendly tail wagging and sniffing ceremony common among neighborhood dogs. Coyotes usually run in pairs, though you often see only one at a time. Many mate for life, some only for the season, and others live in solitude. They also may run in a pack when food, especially carrion, is abundant.

Coyotes invented country-and-western singing long before cowboys rode the range. The Aztecs called them *coyotl*, or barking dog. Even their Latin name, *latrans*, refers to their frequent moonlight serenades. Coyotes' music must be heard to be appreciated. The barking, yodeling song alternately sounds like a lonesome wail and the frenzied greeting of old friends. Once in Big Bend a predawn chorus of yaps and howls sounded like a dozen excited animals. But at first light a single pair came trotting through the campgrounds checking the garbage cans just like city mutts.

A coyote's disharmonious howling, the signature of the Old West, now echoes through the night in every state, including Alaska and most of Canada. In taming the West, humans killed the coyote's major competitors, gray wolves and mountain lions. Without these larger animals to limit its expansion, the adaptable coyote soon spread throughout North America and now ranges south to Costa Rica.

Coyotes seldom receive the credit they deserve. Contrary to Hollywood's Wile E. Coyote, a bumbling idiot who can never catch the roadrunner, these canny canines employ ingenious hunting techniques. Two coyotes may form a relay team to catch a fleet-footed jackrabbit. One rests while the other pursues the rabbit, which invariably runs in a large circle. At the end of each revolution, the fresh partner continues the chase.

Coyotes allegedly cooperate with badgers to catch rabbits in burrows. The badger works digging out the victim while the coyote waits at the exit tunnel. When the trapped animal tries to escape, it lands in the jaws of the coyote. By the time the badger realizes the burrow is empty, Mr. Coyote is long gone with a full belly. What some call cooperation, others call thievery.

Stockmen, like the badger, unintentionally cooperate with coyotes. The predators do not mind at all that ranchers stock the range with animals that have had their defensive behavior bred out. The opportunistic coyote accepts livestock as gifts, or perhaps as tribute, from the humans usurping its domain.

Never a gourmet, a coyote will eat the most easily accessible food. Its diet includes anything edible, plants or animals, dead or alive. Seasonally, berries and fruit (including melons) comprise a large portion of the desert dog's diet. It preys heavily on rodents, rabbits, and ground squirrels in areas where they are abundant. Dead animals, especially along highways, are a standard food item.

Rabbits and mice represent by far the major portion of a coyote's diet. One hundred field mice, about a two-week food budget for a coyote, consume as much forage as a growing lamb. The rabbits on a section of ranchland can easily eat as much grass as six sheep. By helping to control these destructive range animals, coyotes help the rancher.

Ignoring a coyote's contribution to controlling rats, mice, and rabbits on the range, humans have hated and slandered coyotes since the first cowboy rode the range. Naturally ranchers can be expected to turn their fury on any animal that kills their livestock, and coyotes can be expected to prey on sheep and goats. Ranchers traditionally consider coyotes ruthless villains that should be exterminated, but in reality coyotes kill only a small percentage of livestock lost each year. Individual coyotes, however, sometimes do become stock killers and inflict heavy losses.

Coyotes prey more on lambs and angora goat kids than on any other livestock. In 1983, lamb mortality on Texas ranches was 11 percent, with 4.5 percent attributed to coyotes. The kid mortality was 14.8 percent, with coyotes responsible for 5.1 percent. Coyotes killed 57 percent of the total number of lambs and kids lost to predators. Decades of studies show that about twice as many deaths of adult sheep and goats result from disease and other causes as from all predators.

Some control of coyotes inevitably is required in ranching areas. Trapping, poisoning, and shooting may remove problem individuals and reduce local populations. Unfortunately indiscriminate use of poison devastates all meat-eating species, guilty and innocent, harmful and beneficial. Western states yearly expend considerable sums of money attempting to eliminate predators. In 1983, state, federal, and private agencies in Texas spent $4 million to control predators, rodents, and other animals that cause economic damage. Livestock losses amounted to approximately $3.5 million. According to the Texas Animal Damage Control Program, the total benefits of the overall control program exceed the costs by a factor of four.

Despite the concerted efforts of state, federal, and private sectors and the urbanization of its range, the coyote is more widespread now than ever before. The U.S. Wildlife Research Center in Denver reports that almost every major western city has a resident population of coyotes. While practically every other predator has succumbed to the pressures of modern society, the coyote has learned to take advantage of the food habits of civilization. The omnivorous diet and elusive nature of this adaptable canine enable it not only to survive but also to thrive in a modern world.

Deer: Texans' Favorite Game Animal

White-tailed deer
Odocoileus virginianus texanus
Sierra del Carmen white-tailed deer
O. virginianus carminis
Desert mule deer
O. hemionus crooki

Texas is the deer capital of North America. With almost 3.5 million white-tailed deer and 250,000 mule deer, the Lone Star State has the highest deer population in the country. No wonder many Texans take deer hunting seriously. To hunters, the opening day of deer season in mid-November is as traditional as the Super Bowl, Thanksgiving turkey, and the Fourth of July.

With the first nip of autumn, hunters begin to see visions of a mule deer silhouetted against the setting sun with its giant ears straining to pick up the slightest rustle—or of a white-tail with its nose in the breeze and flaglike tail straight up, a sure sign it is about to bound away. This malady common to hunters is called buck fever.

Texans should not brag about the abundant number of deer in the state. Texas has too many deer. Finding enough food poses a real problem for the animals. Most rangeland suffers from overgrazing by sheep, goats, and cattle. Without a protective cover of vegetation, the soil washes away with the rain. Weedy plants with low nutritional value invade the abused areas and become established. After a century of overgrazing, the state's most productive deer habitat, the Edwards Plateau, does not have enough annual protein available for healthy deer. As a result, deer seldom reach their full size, and large numbers starve during dry summers and severe winters. The South Texas Brush Country is the only area in the state with sufficient nutrition for deer consistently to reach their full genetic potential for size and antler development.

For 100 years, ranchers and farmers have killed the wolves, coyotes, and cougars that historically kept the deer from overpopulating. The predators removed the old, sick, and surplus juveniles from the breeding population, leaving only the healthiest animals to reproduce. Despite blinds, high-powered rifles with telescopic sights, and a hunting tradition as old as the state, Texas hunters do not kill enough deer, and they insist on killing the wrong ones. Each year, a half-million hunters take to the woods and kill 300,000 deer, about 150,000 short of the annual harvest required to maintain a healthy deer herd through most of the state.

Since hunting is the only way left today to limit the state's deer herd, hunters need to kill more does. Hunters want trophy bucks with a rack to hang on their den wall, but shooting a buck does little to decrease the population. Most of the bucks killed are inexperienced yearlings, which leaves the mature animals to breed with all the females. Each year about 80 percent of the antlerless deer permits go unclaimed. The deer population will not diminish until the number of females is reduced. On the Edwards Plateau, home of half the deer in Texas, hunters need to kill three times as many does to bring the population in balance with the available food supply. State biologists feel that 20 percent, or 360,000, of the adult does should be removed annually. If hunters killed only does, the deer population would approximate the healthy number for most of the state.

Texas averages 3.5 white-tailed does for every buck, with an average of 1 fawn for 2 does. A healthy female with plenty of food generally conceives twins, and white-tails sometimes have triplets. White-tails usually give birth in May and June, and mule deer from June through August. Food availability, both for the pregnant and nursing female and for the baby after weaning, has the greatest influence on how many fawns survive the first year. Disease and predation, especially from dogs, also take their toll. As many as 75 percent of the fawns may die during stressful years, yet infant mortality alone is not checking overpopulation.

The classic Disney movie *Bambi* has formed an enduring impression for Americans. The helpless condition of the dappled fawns fills us with compassion, sometimes to the point of rescuing "deserted" young. But the fawns are far from deserted or helpless. Their spots camouflage them in the mottled light that filters through the trees as long as the fawns remain motionless, which they instinctively do. The mother feeds only a short distance away and returns several times daily to nurse the young.

When the fawns are two months old, the mother allows the yearlings to rejoin the family. White-tailed deer are moderately gregarious, and family members forage together during most of the year. In areas with abundant food, several family groups may feed together, giving the appearance of a large herd. Mule deer are more gregarious and form herds before, during, and after the breeding season.

As November approaches and restless hunters start showing signs of buck fever, the mature bucks begin suffering from a different kind of fever. Their thoughts turn to the does. White-tailed does come into heat from late October through November, and mule deer about a month later. Their scent attracts bucks like ants to a picnic. Though normally shy and reclusive, bucks become bold and fearless during the rut, or breeding season. With polished antlers they battle saplings and bushes, anticipating actual combat with bucks that may invade their territory.

Deer have small home ranges and seldom wander more than a mile. White-tailed bucks mark their territory by pawing up a patch of ground under an overhanging limb. They urinate on the ground and rub the scent glands near their eyes against the limb. These "scrapes" are sure signs that a buck is in residence. A buck also leaves his calling card by rubbing the musk glands on his hind legs on saplings.

Bucks do not limit themselves to the females in their territory; they will chase the scent of a doe with wanton disregard into the territories of other deer. The clash of antlers echoes through the woods as males spar vigorously with intruders. Hunters take advantage of the bucks' obsession and rattle two old antlers together to sound like a fighting pair. Often the reigning buck in the area comes running to investigate the suspicious sound.

The bucks tend to disregard the dangers of a hostile world during the rut, but the females become even more secretive and solitary. They even drive away their yearling offspring. Despite the many paintings depicting a buck, doe, and fawn, mothers with young rarely tolerate the presence of other deer, including fathers. The males breed with as many females as possible and have no family ties or parental responsibilities.

Unlike cattle, which graze on grasses, deer browse on leafy herbs, fruit, acorns, leaves, and twigs. In the desert, mule deer favor lechuguilla stalks, sotol, mesquite beans, juniper, and leafy plants. Deer consume very little grass, even in times of extreme food shortages. They do not compete with cattle, but they do eat the same foods as sheep and goats.

Dinner time for deer comes just before dawn and again at dusk. Although they usually bed down in a protected location for the day, they may be active at any time. Deer are more active on windless nights when their acute sense of smell can best alert them of danger. They have sharp vision, but like most animals of the night they're color blind. The bright orange vest worn by a hunter looks dull gray to a deer. Despite their acute vision, deer have difficulty discerning the details of stationary objects. Many times I have had a deer approach to within a few yards before it caught my scent and bounded away.

Since deer feed mainly on low-growing plants and have no reason to fear owls or hawks, they seldom look skyward. Hunters take advantage of this characteristic and construct blinds in trees or elevated on poles. These usually overlook a clearing that deer frequently cross or that has been baited with corn.

Mule and white-tailed deer are biologically, ecologically, and behaviorally distinct. The common names hint at the major differences in appearance. Mule deer have large ears, 11 inches long and 6 inches wide, that they cock when alarmed. When the white-tail senses danger, it holds its bushy 10-inch tail straight up. The dark tail has a snow-white underside that flashes like a flag as the deer runs. Mule deer hold their tail down when running. Until a mule deer matures at three or four years, its antlers resemble the white-tail's. At maturity, white-tails have four or five points growing from the main beam without any branching, while each beam of a mule deer's rack divides into two equal branches with two points per branch. Generally mule deer weigh 15 to 20 pounds more than white-tails of the same age.

The desert mule deer of the Trans-Pecos favor a different ecological habitat from the white-tails'. Mule deer live in steep, rugged, open terrain and never inhabit flat prairie and plains country. White-tails prefer heavy brush and dense woodlands. The animals' flight behavior corresponds to their different habitats. In open country, mule deer can keep a watchful eye on anything threatening and will run a safe distance and stop. On the other hand, woodland-dwelling white-tails cannot see what danger lurks behind the bushes, so they do not feel safe until they have put considerable distance between themselves and the threat. A running mule deer has a high, bounding gait necessary to clear boulders and cacti, while a white-tail runs in a low gallop to avoid overhanging branches. A mule deer once was clocked at 38 mph for a short burst, then it slowed to 23 mph.

White-tails live in all but 3 of the contiguous 48 states. This highly adaptable animal is divided into 16 subspecies that range north into Canada and south through Mexico to South America. The Sierra del Carmen white-tail, a mountain subspecies, lives only in the high mountains of the Big Bend area and Mexico. In the rest of the Trans-Pecos, white-tails favor mountain foothills with a heavy cover of thick brush.

Seven subspecies of mule deer live from central Mexico to coastal Alaska. They are found in the semi-arid deserts and mountains of the western United States and Canada and the badlands of the Great Plains. In Texas, desert mule deer live in the Trans-Pecos and the canyons and rugged terrain of the High Plains.

The Dolphin: Delightful Mammal of the Deep

Bottlenose dolphin
Tursiops truncatus

Do dolphins save drowning swimmers? Do they chase away sharks and surf in the waves beside humans? Questions buzzed around the boat as the curious tourists crowded the railing and scanned the waves in anticipation. The captain of the excursion boat had just announced that he had turned on his fish finder to attract dolphins. As though on cue, two dolphins burst through the ship's wake, rode the waves while everybody cheered, then disappeared beneath the waters. Such scenes are frequently reenacted along the Texas coast as dolphins play escort service for boats.

Dolphins, also called porpoises, have thrilled and mystified humans for thousands of years. As opposed to the fish with the same name, dolphins are sea-dwelling mammals. They breathe air, feed their babies milk, live in groups, and interact with humans in ways unexpected for animals. Countless stories recount how dolphins have helped swimmers, towed boats, and played with children. Today trained dolphins at amusement parks add to the mystique. They jump through hoops, walk on water, and propel their 600-pound bodies 16 feet out of the water to pluck fish gracefully from the trainer's hand.

Thirty-two species of the family Delphinidae ply the oceans of the world. Some live in the shallow zone along the continental shelf, while others prefer deep waters. The species found in Texas waters is the bottlenose, the most common dolphin in the Gulf of Mexico and the star performer in oceanaria around the world. Texas dolphins hunt a variety of fish in the shallow bays and channels, but mostly eat striped mullet. Seldom do they venture farther than 12 miles from shore. A 1978–79 survey counted about 2 bottlenose dolphins per mile along the Texas coast, one of the highest populations in the Gulf of Mexico; Aransas Bay had highs of 104 dolphins in October and 281 in January.

Dolphins may thrill us with their tricks, but they fascinate marine scientists for other reasons. A dolphin's skeleton shows traces of hind limbs, and its front flippers have the same bone structure as the human arm. Their close relation to terrestrial mammals indicates that dolphins lived on dry land before they inhabited the sea. Fossils indicate that mammals colonized the ocean about 58 million years ago, probably filling empty niches caused by the mass extinction of reptiles.

Though vastly different from fish, dolphins are equally well adapted to a watery lifestyle. Powerful swimming muscles propel these streamlined animals through the water as fast as 22 mph. Their flexible, torpedo-shaped bodies and oily skin secretions reduce drag and turbulence. Most species feed on fish in shallow waters or near the surface, but some can dive 2000 feet. Human divers must carefully regulate their ascent from even moderate depths to prevent the nitrogen absorbed in their blood from expanding and causing the painful and deadly sickness called the bends. Dolphins never get the bends. When dolphins dive, the water pressure squeezes the air out of their lungs into bronchial passages where no gas exchange occurs. Some species can remain underwater for as long as one hour. Their muscles contain a large amount of myoglobin, which combines with oxygen and stores it for later use. Where is a dolphin's nose? Not at the end of its snout, but on top of its head so it can breathe when it floats on the surface.

Dolphins live about 30 years, and take 5 to 10 years to mature. Like humans, dolphin young have a long learning period. Many of the dolphins seen cavorting alongside ships are mothers with their offspring. The males live in separate groups, or pods. Pods in Texas waters average 5 to 6 dolphins, depending on food availability, among other things. Several pods may join to form a large herd. Dolphins stay in a loosely defined home range, but may switch from pod to pod. As researchers learned about the dolphins' complex social structure, they realized that many of the tales about these friendly animals were based on facts, not imagination.

For dolphins, cooperation is an important facet of survival in the deep. They group around a female giving birth and help the baby to the surface for its first breath. Dolphins with vulnerable young drive away sharks, but if not threatened usually ignore the predators. Sometimes one female will baby-sit the pups while the others hunt. Dolphins often cooperate while hunting. A pod surrounds a school of fish and forces it into a compact cluster; the animals take turns rushing in to feed while the others prevent the fish from scattering. They also corral fish in shallow water. Occasionally, they flush a school onto the shore then swim out of the water to catch the floundering fish. In Africa, fishermen have learned to take advantage of these natural hunting strategies. They wade in the surf and call dolphins by slapping the water. The dolphins herd fish into their nets and get an easy meal in exchange. Texas fishermen report that dolphins respond to the shift in a shrimp boat's engine noise and to the whine of the net winches hauling in a catch. The clever animals follow along and feed on the discarded fish.

Helping fishermen, saving drowning humans, and playing are extensions of a dolphin's natural behavior. Two dolphins will swim on either side of an injured companion, keeping it afloat until they get it to shore. One dolphin even pushed a mattress onto a beach. Dolphins love to play. They spring from the water, race boats, and surf in the waves. The tricks they learn to perform at oceanaria may entertain us, but the tricks they learn by themselves astound us more. A dolphin in a large aquarium saw a diver cleaning the inside glass. The dolphin began to imitate the diver and even made noises and bubbles like the scuba tank. From then on, the dolphin took over the job of cleaning the glass. In another aquarium, a visitor blew a puff of smoke at a baby dolphin. The baby immediately swam to its mother and got a mouthful of milk, then returned to the visitor and puffed the milk at him. The milk made a white cloud in the water like smoke in the air.

Though dolphins have been held in captivity since 1914, scientists did not discover that they used ultrasound for echolocation until 1958. By sending out a stream of 1000 clicks per second, dolphins can read the ocean floor like a road map. This sense is so sophisticated that they can tell the difference between various types of metal, or if a fish has a hook hidden in its body. A research vessel placed a fence of metal pipes across a narrow channel used by a pod of dolphins. The dolphins detected the fence from 1000 feet away. A scout went ahead and investigated, then returned to report to the others. The biologists recorded the dolphins' excited conversation before the animals ventured around the barrier.

While hunting, playing, or traveling, dolphins stay in constant communication with whistles, beeps, belches, and grunts. One experiment tested the ability of dolphins to communicate with each other. Two were trained to get food by pressing a paddle when a light came on. Then they were placed in a tank separated by a curtain with the light on one side and the paddle on the other. The dolphin with the light quickly learned to signal its partner to press the paddle when the light came on. Scientists recorded the conversation, but were unable to decipher it.

Communication, cooperation, and a complex social structure all require a high level of intelligence. A dolphin's brain is larger and has more convolutions than a human's, but has fewer layers of neurons. The ability of dolphins to communicate excites marine scientists, such as Dr. Louis M. Herman at the University of Hawaii. He has trained one dolphin to understand arm gestures and another to understand computer-generated sounds. The animals are forcing us to redefine our beliefs that only humans can comprehend words and sentences. The dolphins understand that words are symbols for objects and that the order of words in a sentence changes the meaning ("take boat to ball" vs. "take ball to boat," for example). Diana Reiss at Marineworld/Africa USA near San Francisco gave dolphins an underwater keyboard. Each of the nine keys has a different meaning and sound. The dolphins press the keys and imitate the sounds when they want an object. Someday this kind of research may develop an artificial language that will enable humans and dolphins to understand each other. If that happens, we may discover that these smiling creatures want to teach us a few tricks.

Foxes: Clever Canines

Red fox
Vulpes vulpes
Gray fox
Urocyon cinereoargenteus
Kit fox
Vulpes macrotis
Swift fox
Vulpes velox

"I was afraid they would kill my cows, so I shot the parents," the gruff-looking farmer said. Then he softened. "But I couldn't kill the pups. Can you take care of them?" Expecting coyotes, I took the box and looked at the five squirming, woolly babies with their eyes barely open. To my surprise, they were red foxes. The misguided farmer had thought that a diminutive fox, not much bigger and no more dangerous than an alley cat, posed a threat to his cattle.

Unfortunately, the fox still carries the tainted reputation that Aesop depicted some 2600 years ago: a clever, cunning animal, but always the villain. We call a person who is a sly rogue a "fox" and say we've been "outfoxed" when we are deceived. Too often, the fox is appreciated only as an animal that provides a sporting chase or has a valuable pelt. Other than that, many consider it vermin.

I grew up with a more positive image of this remarkable animal. Early in the year when the swelling buds promised spring, the red foxes' yapping echoed through the woods surrounding our house, a sign that mating season had arrived. Once, a pair chose the old stone fence in our backyard as a denning site. That ancient wall stemmed the flow of trees and vines trying to reclaim the fertile East Texas soil. It marked the end of manicured lawn and the beginning of wildness. We heard but seldom saw the secretive family of foxes. Only once did I catch a daytime glimpse as a bounding, bushy tail disappeared into the brush.

Like most members of the canine family, all foxes make good parents. Since they cannot dig well, the pair prefers to remodel an abandoned den—a burrow for the red, kit, and swift foxes, but often a rocky crevice or hollow limb high in a tree for a gray fox. The male fox brings the female food while she nurses the litter of three to five pups. The babies open their eyes in 8 to12 days and can walk in 3 weeks. In 6 months the young are fully grown, and the adults and juveniles disperse to claim their own separate territories.

Though the rock fence that served as a fox nursery ended a few feet behind our chicken coop, the wild canines never molested the chickens. As every fox knows, a fat mouse or a tender cottontail makes the best gourmet meal. But foxes are not too picky to dine on chicken, or any other food that comes easily. Foxes are survivors—opportunistic omnivores. An adult red or gray fox must catch about five pounds of food per week. It dines on grasshoppers, beetles, and crickets in the summer, berries and other wild fruit in the spring, earthworms that cover the ground after a rain, and carrion whenever it is available. Despite some hunters' contentions, foxes do not prey on bird nests frequently enough to have an adverse effect on pheasant and quail populations. Mice and rabbits are too easy to catch for a fox to waste time looking for nests. An observer once saw a grouse successfully fledge a nest just a few yards away from a fox's den.

A fox spends its nights, and often part of the day, trotting along at 4 to 5 mph searching for food. Its tireless gait inspired a ballroom dance, the fox trot. While hunting, a fox keeps its ears cocked, listening for the slightest rustle in the grass or the soft squeak of a mouse. Its sensitive nose constantly samples the breeze for the faintest aroma of anything edible. A foraging fox zigzags through fields and may stand up on its hind legs for a clear view through the grass, then almost effortlessly pounce on a mouse, gopher, or field rat.

In a bounding chase, the fox uses its long, bushy tail to help maintain its balance. The furry tail serves another purpose on cold nights—the animal curls it around its face as a nose warmer.

The fox's keen senses of hearing and smell make it an efficient predator of small mammals and equip it well for survival. A rancher in South Texas once encountered a gray fox heading straight toward him. Thinking the animal rabid since it showed no fear, the man shot it. To his surprise, he found the fox perfectly healthy, but blind. Because of its sharp senses, the sightless fox had been able to survive in the unpopulated countryside. Other than humans and automobiles, a fox has only to fear coyotes, bobcats, and an occasional eagle.

The red fox has rightfully earned its stereotypical "foxy" image. Like a riverboat gambler, it is always ready to match its wits with any challenger. Since a red fox cannot win a race with the hounds by stamina alone, it must rely on instinctive and learned survival skills. Herein lies the charm of the chase for the hunter. A red fox will double back or crisscross its trail, run along the top of rock fences, backtrack through streams, and circle behind the dogs. Members of the European nobility donned red coats, had the kennel master fetch their pedigreed hounds and the stable master their horses, and made a day of tea and chasing the wily creature. The fox cooperated by leading the riders and baying dogs for a merry romp through the woods. The noble tradition lives on in America. Backwoodsmen sip home brew and listen in anticipation as the howls of ol' Blue and the other hounds disappear in the distance. Will the fox outfox the dogs?

Fox hunters disdain the unsporty manners of the gray fox. After a short run, it inevitably heads for the closest rocky den or climbs a tree. Unlike any other member of the dog family, gray foxes climb with the agility of a cat. They hook their claws in the bark of the straightest trunk and shimmy right up to the first limb. They may stand in plain view just above the frenzied dogs or hide in the leafy crown of the tree. In any event, the thrill of the chase is too short-lived to satisfy a devotee of the classic fox hunt.

Four species of foxes live in Texas: the red, gray, kit, and swift fox. Each has its distinct habits and habitats. The red fox occurs north of San Antonio, except for the Trans-Pecos and the Panhandle. With its characteristic white-tipped tail, it haunts the edges between forests and meadows, woodlots and fields. A diverse habitat with mixed hardwoods, rolling cropland, brush, and pasture makes an ideal home. In contrast, gray foxes are more shy and retiring and avoid human occupation. They range throughout the state, except for the Panhandle north of Palo Duro Canyon. They prefer dense woods, swamps, thickets, and chaparral—areas with a thick understory of vegetation. The brush country of South Texas and the scrub vegetation and rocky outcrops and canyons of Central and West Texas provide ideal hunting, hiding, and denning sites. While the red fox dines on succulent dewberries, wild grapes, plump meadow mice, and cottontails, the gray fox relishes juniper berries, cactus fruit, mesquite beans, kangaroo rats, and jackrabbits.

No one knows for sure whether the colonists found red foxes in North America or brought them with them, or both. Red foxes were introduced into East and Central Texas for hunting. Fossils found in Pennsylvania indicate the gray fox has called America home for at least 25,000 years. Possibly because of the cold climate of the Ice Age, foxes disappeared from New England and did not return until the last century. Now the gray fox lives throughout the United States, except for the northern Rocky Mountains, and south into Central America. Except for the desert regions, the red fox covers the North American continent, and roams across Europe and Asia as well. Northern red foxes have several color phases: black, silver (black with white guard hairs), cross (dark across the shoulders and back), bastard (bluish gray), and Samson (no guard hairs). Foxes with the silver coloration bring the highest fur prices and are commercially raised.

The two other foxes that live in the Lone Star State, the kit and swift, are much smaller than their cousins and rarer. The kit fox lives in the deserts of the Trans-Pecos, the western United States, and northern Mexico. The swift fox occurs in the grassland prairies that range from the Texas Panhandle north into Canada. The only other fox occurring in North America is the arctic fox, *Alopex lagopus*, an inhabitant of the polar regions.

One night while camping in the desert at Big Bend National Park, I was visited by an inquisitive kit fox. No larger than a cat, but sporting oversized ears, the little critter smelled an empty tuna fish can in my trash sack. Twice it came and prowled around my car while we sat quietly a few yards away. Looking for the source of the enticing aroma, it almost jumped into the car. In hushed wonder we watched the little creature prancing back and forth in the moonlight like a phantom from a world where animals have nothing to fear from humans. Finally, it disappeared into the night to search for a more traditional dinner.

The closely related kit and swift foxes are the smallest, yet fastest, foxes in North America. For 100 yards, they can outrun any prey or predator that they might encounter. Whether in pursuit of a rabbit or trying to escape a coyote, they can change directions at lightning speed. Such an effective defense probably explains why they act so fearlessly and without caution. Lacking the wariness and suspicion of the red and gray foxes, these innocent canines have suffered greatly from coyote and other predator control programs that distribute poisons. As a result, they have been eradicated from large areas in the United States where they once thrived. Fortunately, with the curtailment of predator poisoning campaigns, the swift fox appears to be returning to its original range. Perhaps the kit fox will also return.

Trappers in Texas harvested 50,459 gray foxes in 1986–87 and 7,298 red foxes. At the going rate of twenty dollars per pelt, trapping foxes is more than a million dollar per year industry in Texas. State laws classify foxes as fur-bearing animals and prohibit hunting or trapping without a license. A fox that escapes hunters and trappers usually lives 4 to 5 years, though foxes have survived 15 years in captivity.

The worldwide distribution of the fox has brought it in close contact with humans for untold thousands of years. As it trod through time, this remarkable creature left its imprint on human folklore and language from the Mediterranean to North America. Unlike many wild animals suffering depletion of numbers and loss of habitat, the red fox has benefited from human interaction. Hunters, instead of diminishing its population, have expanded its distribution across three continents and recently transported it overseas to Australia. So, because of its appeal as a game and fur animal, the red fox has managed to extend its range in modern times. A pretty foxy maneuver, huh?

The Javelina: Prickly Pear Gourmand

Javelina, collared peccary
Dicotyles tajacu angulatus;
formerly *Pecari tajacu* (Linnaeus),
Tayassu tajacu, Pecari angulatus

One dark night in the Basin campground of Big Bend National Park, I heard a rustling noise in the bushes near my tent. Critters frequently invade the camp, but the grunting, munching, and lip-smacking sounded like a monster from outer space devouring helpless earthlings. Fortunately the ten eyes that reflected the light of my flashlight belonged instead to five javelinas pigging out in a prickly pear patch. Eventually these four-legged gluttons wandered on to the next gourmet course of their progressive dinner.

Javelinas have the table manners of hogs, and many people confuse them with domestic pigs gone wild. In reality, they are related only distantly to our familiar porkers: both belong to the order of hoofed animals. Fossil records of their ancestors in America date back 22 million years. When the Spanish introduced swine to the New World in the 1500s, javelinas ranged from Argentina to Texas and even into Arkansas.

Javelinas retreated south as ranchers converted most of Northeast Texas into open pastures, but Texas still has more javelinas than any other state. They favor areas with heavy undergrowth and plenty of prickly pears. Much of the state remains prime habitat, especially the Brush Country of South Texas and the deserts of West Texas. Their range extends westward along the southern border of New Mexico into Arizona, and south through Argentina.

These roving garbage disposals eat almost any vegetable matter. Contrary to tales of hunters and ranchers, javelinas do not eat snakes, eggs, birds, rabbits, baby sheep and goats, or any other animal matter. Just give them prickly pears, lechuguillas, sotols, mesquite beans, acorns, and century plants. Since javelinas eat little grass, they do not compete with cattle for food. Prickly pears are essential to javelinas, not only for nutrition, but also for water. They devour the thorny pads and fruit with complete disregard for the spines. Javelinas can go six days without water, and months if their diet includes prickly pears. Three pounds of the cactus a day will keep an adult from becoming dehydrated. Prickly pears are so important that javelinas seldom live in locations void of the plant.

Javelinas have adapted perfectly to life in the hot, arid Southwest. When they are stressed by lack of water, physiological changes reduce water loss from respiration by 68 percent and from urination by 93 percent. Javelinas also adapt their daily activities to conserve water. During the hottest times of the year, these hardy animals forage through the desert only in the cool of the morning, evening, and night. Even in the winter, temperature strongly influences the javelinas' activities. They forage during the warm afternoons, and freezing temperature causes them to huddle together for warmth.

Touching is a way of life for these gregarious animals. Of the 31 behavior patterns observed among herd members, 17 involve touching. Javelinas groom each other more than any other social interaction. Two animals stand close together and rub their noses on each other's hindquarters. Javelinas have a navel-like gland located just above their tail that exudes a musky liquid that helps them identify each other. Grooming and touching probably help maintain the social bond among the individuals of the herd. Besides body language and scent, javelinas communicate with a vocabulary of at least 15 signals. They vocalize to indicate aggression, submission, alarm, and other reactions. A javelina chatters its teeth not from cold, but to threaten its neighbor.

From the viewpoint of a cactus, javelinas are ruthless,

vicious animals. They attack prickly pears without hesitation and voraciously consume the plant. But when alarmed by a human or animal, they scatter in all directions. An inexperienced hunter once froze when a herd of the startled, grunting animals seemed to charge him. One actually brushed his pants leg as it scampered by, unaware of the exact location of the human intruder.

The scatter response, typical of herding animals, confuses the predator and makes attacking a single animal difficult. But humans interpret the flight of fear as the attack of savage animals. After scattering, the nearsighted javelinas may wander about nervously woofing and popping their teeth together, unsure if danger still exists. More than one excitable hunter has sought refuge in a thorny mesquite thinking these timid critters were circling for the kill. Only when defending itself does the javelina bare its two-inch canine teeth and become a serious threat.

The millions of generations of javelinas that have survived through time have developed the senses necessary for survival and lost those that were unnecessary. Javelinas do not need keen vision to search for plants in dense underbrush, but acute hearing and smelling are essential to their social life. Their complex vocabulary keeps the herd close together while foraging through heavy cover, and individuals recognize other herd members by scent, a necessity for social animals.

An equal number of males and females make up the herd, but only the dominant male breeds with the females. Pregnancy lasts for 145 days, and twins are common. The young can walk and follow their mother after the first day, and mature in about 48 weeks. Adults weigh from 25 to 60 pounds, stand 20 to 23 inches tall, and are 34 to 38 inches long. The size of an animal, more than its sex, determines its place in the pecking order.

In Texas, herds average from 15 to 17 animals, depending on the amount of food available. Herds remain stable, but they often have several subgroups of about 6 that may eventually break off to form new herds. All members defend the territory—about 400 to 585 acres—against intruding javelinas. The animal's short legs and stout body were not built for travel, but for scampering through the underbrush. A javelina seldom ranges more than a square mile during its decade-long life.

Even though the javelina was elevated to a game animal in 1939, Texas hunters and ranchers have traditionally considered it vermin. Ranchers dislike this not-so-innocent critter because it tends to slash up the cow dogs that search thickets for stray cattle. Hunters sometimes have the same problem with their bird dogs. Javelinas have a natural animosity for dogs, probably because coyotes kill javelina young. They do not have tusks like wild boars, but their two-inch canines can inflict a serious wound. Any cow dog or bird dog attacking a javelina usually requires an emergency trip to the veterinarian for stitches.

Today many stockmen value javelinas primarily because they can help increase profits. By eating prodigious amounts of prickly pear, the animals do more to control cactus than many range-improvement programs. Many ranchers who sell hunting leases now view these creatures as a money-maker along with deer. But whether or not the javelina has the approval of the ranchers, this animal has a rightful place on the range along with the more appreciated denizens of the Southwest.

The Longhorn: The First King of Texas

When the Civil War broke out, almost every man of fighting age left Texas to defend the honor of the South. As he was leaving, many a father told his oldest boy, "Son, find a cow with a calf and tie it up in the corral till it's tame enough to milk. Then mother will always have milk for the children while I'm gone." The cow that the young lad dutifully chased out of the thorny thickets, lassoed, and tied to a post in the family corral was the wild longhorn. A longhorn, especially a mother with a calf, would just as soon gore a horse and rider as swat pesky flies with its bushy tail.

"Texas was born of hide and horn," goes an old frontier poem. The hide and horn belonged to what settlers considered the toughest, meanest, sliest, most independent critter that ever called Texas home—the longhorn. These ornery bovines became legendary long before they became cash on the hoof at Kansas railheads. Big game hunters from Europe considered them the fiercest animals in North America. With six-to-nine-foot horns and a disposition that made a bear seem mellow, longhorns challenged the bravest hunters from the continent.

After the war, the returning men found their state bankrupt and in the depths of a depression. The only salable commodities were the hides and tallow of the longhorns that had run wild for centuries. Like bison on the prairies, the cattle were shot and skinned and their carcasses left to draw flies. Then the railroad reached Kansas and made the hungry East only a trail drive away. A $3 cow in Texas brought $40 in Abilene, Kansas. Large ranches formed in South Texas, and cattle became king of the economy. With so many men lost in the war, ranchers had to employ adolescents to herd the cattle, giving birth to the term "cowboys."

The cattle that blessed Texas bankrolls were gifts from the Spanish conquistadors and padres. Christopher Columbus brought the first cattle to the New World on his second voyage, and in 1540 Coronado, searching for the fabled golden cities of Cíbola, took the first herd north of the Rio Grande. By 1600, hundreds of thousands roamed free in Mexico. Spanish missionaries brought the first breeding cattle into East Texas in 1690. By 1770, a herd of 40,000 grazed the grassy plains that surrounded the Mission Espíritu Santo. The mission is now a state park at Goliad. Unlike bison, longhorns refused to graze in large herds across the prairies. They preferred the protection of impenetrable thickets and cedar brakes. During 200 years of running wild, a breed of rangewise cattle that Texas could call its own had developed.

As longhorns spread across the state, they had to survive disease, drought, scorching heat, and freezing weather. In the winter, they pawed through ice for food and browsed bushes like deer. Besides the natural calamities, wolves, cougars, and other predators culled out the weak. Only the hardiest, fleetest, most resourceful creatures lived to produce offspring. Like a potter molding clay, natural selection shaped a breed with a survival instinct that no improved breed can match. The endurance and stamina of the longhorn made possible the perilous trail drives destined for the future.

By the mid-1800s, millions of longhorns hid in the chaparral country north of the Rio Grande. They ranged north to the Red River, east to Louisiana, and west to the Brazos River. These cautious animals hid in the brush by day, grazed by night, and would flee at the first sign of a rider. Early settlers tried in vain to tame them, but even the captive calves grew into outlaw cattle. Longhorns, though numerous, proved so difficult to kill that in times of famine the pioneers resorted to killing mustangs, which also ran wild across the state.

The wild longhorns bore no brands, and any rancher could round them up and burn his sign on their flanks. The first cowboys in South Texas had to "pop the brush" to drive out the reluctant longhorns. Their rugged ponies, themselves products of the brush country, bounded through the thorny growth like a bronco at a rodeo. The rider dodged and ducked for his life, fighting to stay on his horse while he roped and threw the cow. Without his hat, jacket, leather leggings, boots, and gloves, the South Texas cowboy would have been skinned alive by the thorny vegetation. Later, cowhands on the vast ranches in the Panhandle had the luxury of chasing cattle across open prairies, unimpeded by brush or thorn.

Both the longhorn and the cowboy adapted to living in the wildest part of Texas by becoming tougher than the rugged, unforgiving environment. Although the cowboys proved tougher, it was not by much, and not all the time. Both became legends, and rightly so, because pushing a herd of longhorns 800 miles up the Chisholm Trail had to be one of the hardest jobs ever demanded of man. The cowboys managed the greatest human-controlled migration of animals in history. During the 1870s and 1880s, they "trailed" ten million head of cattle to the northern railheads.

The era of the cattle drive brought wealth to Texas and gave the world a rich and enduring legend, but it lasted only a brief 25 years. By the 1890s, railroads spanned the Lone Star State, and barbed wire dissected the open ranges. Though at first bitterly opposed to fencing the range, ranchers found that confining cattle allowed them to breed more profitable stock. The very characteristics that had made longhorns king of the open range almost doomed them to extinction.

By the turn of the century, cattlemen wanted beef, not brawn. They wanted uniform animals that would fatten up at a predictable rate. An even disposition in a fenced pasture mattered more than survival in the wild. Longhorns had a natural independence and an uncanny ability for eluding fences. A 1500-pound steer could leap like a deer, bust through a fence, and even swim underwater to get into the next pasture. Stories abound of freedom-minded outlaw steers that dropped dead at the sight of a cattle pen, or that lay down and died before being dragged into a herd. So, after 25 years of stardom, exit the longhorn and enter the Hereford and other more manageable breeds. The next quarter of a century saw the fabled longhorn fade like a western sunset.

In West Texas, a few old mossy-horned steers survived in the rugged Glass Mountains after ranchers had converted to Herefords. Windmills dotted the arid Big Bend country, and high-class Herefords roamed the range. The longhorns would come down to the water tanks and gather a herd of a thousand yearling Herefords. At the first sign of a rider, the longhorn steer, as fleet as a coyote, would lead his herd in a mad stampede for the hills. The cowboys spent days rounding up their stock. In time, the young Herefords became lean and wild like the longhorns, and the devil to manage. Finally the frustrated foreman of the ranch mounted his fastest horse and hunted down the last of the noble longhorns like vermin.

By 1920, only a few hundred survived in the remote areas of the Brush Country. Western author J. Frank Dobie, an old South Texas rancher himself, loved the lore of the West and began assembling a herd of the rare animals. Dobie's longhorns eventually became the official state herd, which now numbers about 125 and is kept at Fort Griffin and other state parks. In 1927, Texas congressmen introduced a bill that established a herd in Oklahoma's Wichita Mountains National Wildlife Refuge. Officials inspected 30,000 Texas cattle but found only 20 cows, 3 bulls, 4 calves, and 3 steers that they considered purebred longhorns.

Two hundred years of life in the brush molded other distinctive characteristics into this vanishing breed besides horn size. The longhorns' heavy forequarters make their front legs appear shorter than the rear. Their horns crown a long, wide head topped with a thatch of hair. A lean look, swayback, and hump just above the tail give these rangy critters an unsightly appearance. Steers have the longest horns, followed by cows and bulls. The wide color variation of a longhorn makes every individual unique. The technicolor tones range from red to blue and black to white in abstract splashes, speckles, streaks, and spots.

When the Longhorn Breeder's Association formed in 1964, about 1500 head existed. Now 62,000 graze in state, federal, and private herds. Among the 2500 registered breeders, individuals pay as much as $60,000 for a premium cow. Cattlemen buy longhorns for more than nostalgia, show animals, or investments. With the price of ranching steadily increasing, an animal completely self-sufficient on range grass and one that can survive the hardships of disease and climate represents a genetic gold mine. The fertile longhorn cows reproduce for 25 years, while other breeds stop calving after 8 years. Stockmen are attempting to breed the survival characteristics of the longhorn into the better beef-producing breeds. So the legendary longhorn is assured of a lasting place in Texas, not just in relic herds but also in the genes of future breeds of premium cattle.

The Opossum: America's Strangest Animal

Opossum
Didelphis virginiana

"It has a head like a Swine, a taile like a Rat, and the bignesse of a Cat." The bizarre little animal we call the opossum astounded Captain John Smith and the other Pilgrims of the early 1600s. But long before the colonists set foot on the new continent, the opossum had made its mark on Europe. The Spanish explorer Vicente Yáñez Pinzón discovered the strange "monkey fox" in about 1500. The conquistador presented the lowly opossum to the royal court of Spain, where Queen Isabella slipped her jeweled fingers into the animal's marsupial pouch and marveled at the extraordinary creature from the New World.

The opossum, or possum as it is usually called, baffled the first explorers with more than its strange looks. The more they became acquainted with the opossum, the more they considered it one of nature's strangest jokes. It grins and salivates like an idiot, but has an IQ higher than a dog's. It has opposable thumbs like a human, yet hangs from limbs by its tail like a monkey. It carries its babies in a pouch like a kangaroo until they are old enough to cling to its back or hang from its tail. It has poor vision and cannot run or climb fast enough to escape a predator. Its most famous peculiarity is its tendency to play dead when frightened. These characteristics ensured it a prominent place in southern folklore.

Since the days of the Pilgrims, the opossum has gained more recognition than it would have liked. The 1644 botanist Georg Marcgrave described opossum tail soaked in water as a "wonderful remedy. . . . in all of New Spain there is not to be found another remedy so useful in so many cases." Possum tail reportedly "cured inflammation of the kidneys, constipation, coughing, and if placed on a wound would draw out any ailment of flesh and bone."

The opossum's culinary reputation, however, soon eclipsed its medicinal contributions to society. Possum and taters became a meal time favorite in the South. (Not knowing that "taters" in this case means yams is a sure sign of Yankee ancestry!) Late-night hunting trips with baying hounds and moonshine whiskey became a southern tradition.

The sex life of the possum has aroused more speculation than a soap opera. According to widespread rumors, the male possum copulates with the female's nose and she blows the fetuses into her pouch. The possum's unusual reproductive organs account for this preposterous misunderstanding. The male's penis has two heads to match the twin canals of the female's uterus. The unusual anatomy gave the opossum family its scientific name, Didelphidae, Greek for "double womb."

The female reinforces the myth by her activities just before giving birth. She sticks her nose in her pouch, not to blow in the babies but to lick a path to help guide the young from vagina to pouch at birth. The half-inch babies are born after only 12 days, the shortest gestation of any North American mammal. At birth the babies, hardly more than embryos, weigh only 1/175 of an ounce. An entire litter of 10 to 20 can fit in a teaspoon. The tiny young have strong front legs equipped with claws for crawling the 3 inches to the pouch, but they lose the claws soon after attaching to a nipple. Safe and secure in the pouch, as many as 13 begin the slow process of maturing.

At two months, the mouse-size young begin leaving the crowded pouch between meals to climb onto their mother's back. They cling tenaciously to her fur as she ambles on her nightly foraging trips. Weaned by three months, the juveniles may stay with the mother for another month, and they reach sexual maturity in eight months. Possums usually have two litters per year.

The possum's naked tail has inspired numerous folktales. According to legend, a possum stealing corn near a cemetery saw a ghost. The poor critter was so frightened that all the hair fell off its tail. Another misleading account credits fire with causing the hairless condition. The stories vary, but no one disputes the fact that possums make good use of their tail. Besides helping them balance when running and climbing, they use the prehensile organ as an extra hand to cling to limbs or to carry nesting material.

Possums belong to the ancient order of marsupials, animals with a pouch in which the young develop. While most marsupials occur in Australia, the possum has lived in North America for over one million years. No other marsupial lives north of Mexico, but 65 different species live in Central and South America.

Until the early 1900s, possums ranged north to Pennsylvania, Ohio, Indiana, Illinois, and Iowa. Since then they have moved into Vermont and southern Canada. Extreme winters limited their northern range expansion until human settlements provided warm dens and winter food. Because these nocturnal creatures eat almost anything, including carrion, they do not mind having humans as neighbors. With dog food and garbage providing gourmet dining at night, possums feel as at home in exclusive neighborhoods as in the deepest woods. In Texas, possums favor deciduous woodlands, but they occur in almost every county. In the prairies and arid West Texas, they live along wooded streams and rivers. The residents of Upshur County in East Texas call their county the Possum Capital of the World. If you happen through the small town of Rhonesboro, stop by the International Possum Museum, housed in the general store.

A possum makes its home in a hollow tree, in an empty burrow, or under a house. For a soft, warm den, it bundles up leaves and carries them in a downward curl of its tail. To protect its naked tail and ears, a possum avoids extreme cold. Even in Central Texas it often suffers from a frostbitten tail and ears. In the winter it packs its den entrance tightly with leaves and sleeps through the cold spells. But it will not hibernate; it will emerge as soon as the weather warms.

A possum does not need much space in either a wild or urban setting. Its home range varies from 4 acres in rich woodlands to 50 acres in habitats with less abundant food. It lives alone and wanders only far enough to fill its belly. Two studies in East Texas reported that about 70 percent of a possum's diet consists of insects, worms, and plants. Mammals, primarily cottontails and rodents, accounted for about 10 percent of its food, and reptiles for about 7 percent. In the fall, possums like to eat persimmons and the red fruit of the holly *Ilex decidua*, commonly called possum haw because of its popularity with the animal.

Possums view snakes from a totally different perspective from most other animals. Instead of fear, they exhibit an innate desire to eat them. They eagerly devour rattlesnakes, copperheads, and cottonmouths with no ill effects from even repeated bites. Possums injected with venom from North American and foreign pit vipers, including cobras, showed no tissue reaction or organ damage. Though possums are immune to the venom of pit vipers, the neurotoxin from a coral snake bite is deadly for them.

Possums have neither sharp claws nor speedy legs to protect them from predators, but no animal is defenseless. By human standards, possums look stupid; learning and discrimination tests, however, indicate they are smarter than man's best friend, the dog. The marsupials may appear inept, but they have a few tricks to get them out of tough spots. A threatened possum snarls with lips curled back exposing all 50 teeth: it has more teeth than any other American mammal and no compunction about biting the hand that offends it. Angry hissing and copious salivating accompany the show of teeth, an act that discourages most predators. I once saw a possum stand its ground against a 75-pound German shepherd, who finally gave in and let the snarling beast scamper away. Although the hissing, salivating animals may appear rabid, possums seldom contract rabies.

If the ferocious bluff fails, these animals have one final, and famous, defense: feigning death. They roll over and go limp, with their tongue hanging out and eyes closed. They often discharge a greenish, foul-smelling liquid that helps repel an attacker. "Playing possum" is not a clever trick, but an innate survival response. The animal enters a catatonic state triggered by a brain reaction to fear. It may recover rapidly, but can return immediately to its deathlike trance if danger still threatens.

A possum's bluffing works well. This defenseless creature seldom becomes the meal for its major enemies: dogs, coyotes, bobcats, and great horned owls. Besides human hunters, the other serious threat to the possum is that four-wheeled predator, the automobile. Possums have survived a million years of climatic and environmental changes and have even adapted to urban life, but modern-day traffic may be this enduring animal's greatest test for survival.

The Pronghorn: Home on the Range

Pronghorn
Antilocapra americana

"Where the deer and the antelope play," sang the cowboys as they rode across the broad open range. The antelope in the lyrics was the pronghorn, a native of the rolling prairies and brushlands from Canada to Mexico. Before the cowboys' time, 500,000 pronghorns roamed the prairies with the bison. Settlers romanticized pronghorns in song and legend, but eradicated them and the bison from most of their range. The creatures suffered from massive slaughter by hunters and found their range crisscrossed with barbed-wire fences. Unlike deer, pronghorns never learned to jump even a three-foot fence. Fences block their movement unless they can crawl underneath. By the 1920s, only about 12,000 of these free-spirited animals remained.

Most ranchers had the mistaken idea that pronghorns competed with cattle for food. Any animal that cut profits became the enemy, preferably viewed through the sights of a high-powered rifle. Stockmen hunted and trapped wolves, mountain lions, coyotes, and pronghorns to extinction where possible. Eventually ranchers discovered that pronghorns do not usurp the cattle's grasses but primarily eat shrubby plants and succulent forbs (herbaceous plants other than grass). Once pronghorns gained the favor of ranchers and the protection of state laws, their numbers began to increase. Most of their ancestral range has been farmed, ranched, and overgrazed, but state wildlife departments have reintroduced this magnificent animal wherever possible.

Living almost a million years on the open prairies has adapted pronghorns perfectly to their expansive habitat. For safety, they depend on sight, speed, and their herding instinct. Their large eyes can detect the slightest movement across the treeless landscape. The location of their eyes on the sides of their heads provides a wide field of view, which makes sneaking up on them almost impossible. If a predator does surprise them, they can bound away faster than any other North American mammal.

A pronghorn's body is an engineering marvel designed for speed and endurance in the roughest terrain. Pronghorns gulp air while running; their hearts, lungs, and windpipes are much larger than those of other animals their size. The deep breathing enables the heart and lungs to circulate oxygen-rich blood to the straining muscles. These graceful animals can run at speeds approaching 55 mph.

A thick, cartilaginous pad on their hooves prevents lameness when bounding among rocks and boulders. The tendons, ligaments, and leg bones fit so well together that these rugged animals seldom suffer a sprain. Their front hooves, which hit the ground with the most force, are a half-inch larger than the rear hooves, another adaptation to increase surefootedness.

Herding behavior also helps protect pronghorns in their short grass, shrubby environment. These alert animals constantly look up to scan the horizon while grazing. Many eyes can detect danger faster than a single pair, and the first one sighting a potential threat signals the rest of the herd. If danger is imminent, the herd bolts but stays close together, a typical escape pattern for plains-dwelling herbivores. The movement of the herd confuses a predator and makes selecting an individual victim more difficult—the herd can run together at speeds of 47 mph.

Pronghorns use 31 behavior signals in communicating with each other. A puzzled or suspicious pronghorn raises the tan hair on its back and the black hair of its mane. It walks or trots stiff-legged. If danger threatens, the animal flares the long hairs on its large white rump and dashes away.

Equipped with an unusual curiosity, pronghorns will

investigate any small moving object that they do not recognize. Hunters through the ages have taken advantage of this by tying a rag to a bush or lying down and waving a handkerchief. Often the inquisitive animal can be lured within range, whether it is an Indian's bow, photographer's camera, or hunter's rifle. Texas now has enough pronghorns to have a hunting season. Each year approximately 800 are killed from almost 20,000 in the state.

When you drive through the western half of the state, look for pronghorns grazing across the countryside or near a water tank. Common, though scattered, they range from the Panhandle to South Texas. They usually stay far from the road, but I once saw a herd of ten adults and juveniles grazing near a fence. As soon as the car stopped, the animals nervously stared in my direction and hesitantly pranced away. Finally the dominant male decided the threat was too great and began to herd the others in a forced retreat. He ran back and forth until all had run a safe distance. With their protective coloration, the agile animals seemed to disappear into their surroundings.

From March through the breeding season, from midsummer to October in Texas, a male establishes his territory and collects a herd of from 2 to 14 females. The buck vigorously defends his territory and harem from intruding males and greets trespassers with a hard stare and a snorting and wheezing warning. If the challenger does not retreat, the dominant male chases him; horn clashing and combat, however, rarely occur.

The females have their babies in February and March in South Texas and from May to June in West Texas. They usually have twins and nurse them in separate locations for protection against predators. Within a week, the youngsters can run about 20 mph. The fawns group together in nursery herds, and the mothers return several times daily to nurse their offspring. By the time the male juveniles are weaned in September and the females in November, the young males have established a lifelong hierarchy within their age group.

The young of all mammals like to play, but pronghorns enjoy playing all their lives. Their favorite game is running. They race, charge, circle, and chase each other at the least provocation. Racing is such a thrill that they seem to delight in challenging automobiles and even trains, and frequently dart in front of a vehicle once they pass it.

All this running, both for play and defense, takes a lot of energy, and pronghorns eat literally day and night. They eat at dawn, rest, have lunch, rest during the heat of the day, eat until dark, rest, and continue the cycle through the night. With stomachs half the size of a sheep's, they must eat highly nutritious foods. Succulent forbs comprise about 60 percent of their diet, shrubs about 25 percent, and grasses about 15 percent. Flowers and fruit seldom escape their attention. Pronghorns, perfectly adapted to the food in their habitat, eat large quantities of locoweed and other plants that can kill domestic livestock. Their liver and kidneys, larger than those of domestic animals, can metabolize the toxic elements in the plants.

The horns of this unusual creature make it unique in the animal kingdom. Pronghorns have true horns, like cattle, made of a hairlike structure covering a bony sheath. But unlike other horned animals, the pronghorn sheds its horns annually, the way a deer sheds its antlers. The horns end in a slight hook curving backward. A prong, which gives the animal its name, faces forward. Both males and females have horns, but the females do not have prongs. The males' horns begin growing at two months of age, but the females' delay until their second year and reach about 10 inches in length. The record trophy, a buck with 20-inch horns, was killed in 1899. Today hunters consider 12-to-14-inch horns a good trophy.

Rabbits: Hippity-hopping across Texas

Black-tailed jackrabbit, California jackrabbit
Lepus californicus
Eastern cottontail
Sylvilagus floridanus
Swamp rabbit
S. aquaticus
Desert cottontail, Audubon cottontail
S. audubonii

One chilly spring afternoon, I stopped for gas at the only store in Dew, Texas, population 71, and got a fill-up, a candy bar, and two unusual passengers. The store was the old-fashioned type—the only kind that could survive in this dot-on-the map town since Interstate 10 had diverted the Houston–Dallas traffic ten miles west. The superheated interior from a big space heater and the cold stares of two elderly domino players greeted me as I pushed open the rickety door. My entrance interrupted not only their game but also the conversation between a young girl holding an old cigar box and the lady behind the candy counter. The girl had something, apparently alive, in the box and appeared puzzled about what to do.

"Anything else?" asked the proprietor as she came over to the cash register. My curiosity would not let me leave without knowing what was in the box.

"Maybe a candy bar," I said, and nonchalantly stepped over to the candy counter. To my surprise, the girl had two baby cottontails not more than a few days old nestled in the cigar box. "Where did you get those?" I inquired as I looked at the squirming, almost naked bunnies. "They hardly have their eyes open!"

"My dog killed their mother, and I don't know what to

do with them." A few minutes later, I left the store, got into my car, and handed my surprised wife an Almond Joy and a box of bunnies. We named one Dew and the other Tex, after their birthplace.

For the next week, our lives revolved around holding our two babies in the palm of one hand and feeding them milk with an eye dropper. Rabbits grow fast, and within two weeks they refused to stay in their box. Dew and Tex had an insatiable curiosity and explored every nook, corner, and closet. They jumped on our bed in the middle of the night, got trapped inside a trash basket, and left their pellet-size calling cards in our shoes, on chairs, and under furniture.

Nights, the active time for rabbits, became a real adventure around our house. No more quiet evenings. Our guests bounded through the rooms chasing each other, zigzagging, and jumping from one hiding place to another. We discovered the importance of their play when we tried to catch them. They had perfected the evasive maneuvers necessary for escape and knew the exact locations of all the hiding places. Six weeks after our providential stop in Dew, I released our two wards in some nearby woods.

Rabbits are ingrained in the culture and folklore of our society. When the lights go out after an Uncle Remus story of Br'er Rabbit, children cuddle their stuffed bunny for warmth and security. Peter Rabbit, with his narrow escapes in Mr. McGregor's garden, and Bugs Bunny, with his adventures with Elmer Fudd, are among a child's first folk heroes. The Easter Rabbit even shares the glory of one of Christianity's most revered holy days.

The mythical creatures characterized by Hollywood and commercialized by Madison Avenue exhibit little of the true nature of cottontails and jackrabbits. Far from being witty, brave, and aggressive, rabbits crouch in a protected hiding place by day and venture out to feed only as the day fades into evening. But to survive in what to them is a hostile world, rabbits must be alert, cautious, and quick to respond. Besides human hunters, their primary threat, rabbits have to contend with hungry hawks, owls, eagles, coyotes, foxes, bobcats, raccoons, dogs, cats, and rattlesnakes. So many animals depend on bunnies for food that only one out of six adult rabbits lives to see its second year.

Rabbits cope with such extreme predation in a very simple way—they have lots of babies. In South Texas they breed year-round, and in other parts of the state they have a 9-to-10-month breeding season. A short gestation period, 27 days for eastern cottontails and 41 for jackrabbits, and only 1 to 2 weeks of postnatal care allow every mature female to have as many as seven litters per year. An average litter consists of two or three young, though five to eight are not unusual.

Eastern and desert cottontails dig five-to-ten-inch burrows and line their elaborate nests with soft fur. Swamp rabbits construct their nests above ground, concealed in a tunnel of dead weeds with a side entrance. The babies, helpless and almost naked at birth, open their eyes in about 7 days and leave the nest in 12 to 14 days. Juveniles mature sexually in 4 to 6 months, so those born in the early spring sometimes breed before their first winter. Adult females average about ten surviving babies per year.

Jackrabbits, members of the hare family, do not need the cozy nests necessary for cottontails. Newborn hares enter the world fully furred, with open eyes, and frisky as fleas. They leave the nest, a shallow depression under a bush or other protection, within 24 hours. After a week, the litter scatters and the young are on their own. They mature in about 8 months. Approximately 14 young for every jackrabbit female survive each year.

Rabbits as a species depend on a high birthrate to prevent extinction, but the individual rabbit relies on sharp hearing and keen vision to spot danger, and speed to escape. At the first sign of danger, a rabbit perks up its ears and scans like a radar antenna to detect the slightest sound. The rabbit cautiously hops to safety if the threat is far away. If danger is near, the rabbit drops its large ears, squats close to the ground, freezes, and relies on its brown color to help it blend in with its background. If the predator approaches too closely, the rabbit darts away with evasive zigzags and jumps. Cottontails can sprint 15 to 20 mph, while jackrabbits reach 35 mph and cover six to nine feet with a single jump.

Predators probably exert the greatest influence on the size of local rabbit populations, but land use determines whether a local population survives or disappears. In the last 50 years, habitat destruction by agriculture and ranching has greatly reduced the number of almost all wild animals, including rabbits. But rabbits can adapt to such a wide variety of habitats that they remain one of our most common wild animals.

Grasslands and fields interspersed with brushy refuges provide the best habitat for cottontails. These cautious animals seldom venture far from cover. The farming practices of the early pioneers, with wood lots and brushy fence rows, ideally suited the cottontails, and their numbers increased significantly. But when large-scale, intensive farming cleared vast acreages of pastureland, woodlots, and thickets, cottontails had no safe refuge, and their numbers diminished.

The eastern cottontail is the most widespread rabbit, living from southern Canada southward beyond Mexico. Its range includes the eastern United States from the Atlantic to the Rockies and into New Mexico and Arizona. In Texas, this cottontail occurs in every county except the extreme western counties of the Trans-Pecos.

The swamp rabbit favors floodplains, bottomlands, marshes, swamps, and estuaries, and it seldom ventures further than 1.5 miles from a major body of water. Its numbers are decreasing due to the draining of lowlands, clear cutting of forests, and flooding of riparian woodlands by impounded rivers. Its range extends through the southeastern United States and into Texas west to San Antonio and south to Aransas Bay.

The desert cottontail lives in the arid Southwest, where it favors brushy ravines, thickets, and brambles. It lives in deserts, dry mountain slopes, creosote flats, and arid shrublands from the high deserts of Montana south through the western half of Texas and into Mexico. It is the only rabbit known to climb sloping trees and bushes, and is so common in prairie dog towns that it is sometimes called the "prairie dog rabbit." In Texas, its range is west of a line from Wichita Falls to Brownsville.

The black-tailed jackrabbit bounds across the arid regions of the Southwest and inhabits Texas west of a line from Houston to Dallas. It prefers open, short grass areas where its keen vision and hearing are unimpaired. Jackrabbits thrive in desert shrublands, pastures, and cultivated areas. They survive extreme desert temperatures by crouching in the shade all day and by radiating body heat through the many blood vessels in their large ears. Also, their body temperature can rise between five and seven degrees F during the day without ill effects. At night, the stored heat is released and their temperature returns to normal.

Besides their high reproductive rate, the rabbits' cosmopolitan diet helps them thrive in widespread areas. During the growing season, they eat grasses and almost any herbaceous or succulent plant. In the winter, they survive on shrubs, bark, twigs, and woody plants. In large numbers, rabbits can have a significant impact on crops and range grasses. According to one report, 148 jackrabbits eat as much as 1 cow, or 6 as much as 1 sheep.

A contented rabbit requires only a safe refuge for hiding and a patch of grass for dinner. With these conditions satisfied, it has no reason to travel. In prime habitats, the home range of a cottontail may be less than 5 acres, while a jackrabbit requires about 40 acres. The size of a home range varies according to the abundance of food and cover. Several individual ranges may overlap, with many rabbits congregating on roadsides, in cultivated fields, or in other feeding areas. Rabbits follow well-beaten trails and know the exact location of every hiding place within their home range. When frightened, a running rabbit tends to circle within its range, a habit that hunters, both human and coyote, use to their advantage.

During the breeding season, the males travel freely, while the females remain in their home range and drive away other females. Males develop a well-defined hierarchy, with only the dominant individuals mating. After the breeding season, the juveniles disperse to establish their own home ranges. The broad dispersal of the young, their high reproduction rate, adaptability to a variety of habitats, and a universal diet enable rabbits to be one of the most abundant yet most preyed upon animals in North America.

The Raccoon: The Masked Bandit

Raccoon
Procyon lotor

Almost everyone who spends much time outdoors encounters a particularly brazen animal that wears a mask like a bandit—the raccoon. The dark eye patches and banded tail are the trademarks of this nocturnal prowler. The midnight clamor of upset trash cans announces the raccoon's presence, and the unerring stare of its eyes, reflecting red or green in a flashlight, attests to its boldness.

Raccoons play their role as lords of the night with gusto, and they reluctantly give up the part when the curtain of dawn ascends. Early risers may see a raccoon busily foraging along a stream bank, probing under rocks in the water searching for its favorite food, crayfish. Bits of colorful shells accompanied by long-toed tracks in the mud show where a raccoon enjoyed its dinner.

Though ranging from central Canada through Mexico, and equally at home in deserts, forests, farmlands, and suburban areas, raccoons seldom stray far from water. Aquatic invertebrates comprise an important year-round portion of their food. In East Texas, half of their annual diet consists of crayfish and acorns, supplemented by plants, insects, and rodents. Berries and fleshy fruit take them through summer and fall, and acorns sustain them through the winter. Wild plants make up most of their diet, but the opportunistic feeders never overlook carrion, garbage, corn, orchard fruit, or any other easily obtainable food.

These fearless foragers are famous for washing their

food. Captive raccoons invariably dunk their meal and then play around with it in their water dish. I once saw a raccoon in a zoo try to wash a snowball. This stereotyped behavior gave them their Latin name, *lotor,* meaning "the washer." Raccoons in the wild, however, show no compunction about eating either dry or dirty food. The food washing of captives is merely a substitute for their normal aquatic feeding.

As winter approaches, raccoons become butterballs, and their waddling gait only adds to their obese appearance. A raccoon gorges all summer on berries, grapes, persimmons, wild plums, crayfish, and whatever corn, grains, rodents, birds, and fish it can find. By fall, it may have doubled its weight, but unlike that of humans, the raccoon's fat serves an important purpose. When food becomes scarce in the winter months, the stored body fat tides the animal through until the next spring. In northern latitudes, where deep snow covers the ground all winter, raccoons may sleep for four months. Unlike true hibernating animals, though, a raccoon's metabolism and body temperature remain normal, and the animal wakes up anytime the temperature warms. In the Southwest, raccoons remain active throughout the year and become dormant only during protracted cold spells.

A raccoon's varied diet and widespread distribution hint at its intellect. A highly intelligent and keenly discerning animal can take advantage of a greater choice of habitats and food sources. To survive in new situations, an animal must learn rapidly what is safe and what is dangerous. Laboratory tests place raccoons halfway between cats and rhesus monkeys in speed of learning, and equal to cats in visual discrimination.

Raccoons have keen vision, both during the day and at night when they are most active. They have sensitive hearing and pay particular attention to unusual noises. A highly developed sense of touch and long fingers on their forepaws give them remarkable dexterity. They can catch flying insects and darting crayfish and can pick up minute morsels. In tests, raccoons do as well as humans in distinguishing between objects of different sizes and textures.

Anyone with a pet raccoon soon discovers that its alert, curious nature and manual dexterity often spell trouble in the home. After a young raccoon passes through the cuddly stage, common to all mammal babies, its insatiable curiosity drives it to explore cabinets and empty boxes, and to spread out sewing kits, jewelry, and tools all over the floor. Wild juveniles are equally inquisitive, with object manipulation occupying a major part of their play.

When a pet raccoon reaches adulthood, it becomes more than a nuisance—it can be a terror and a threat to human safety. Companionship plays an important role in the learning and development of babies and juveniles, but they become solitary creatures when they mature during their second year. Solitary animals never develop the social behavioral restraints that govern gregarious animals like dogs. As a result, a pet raccoon may tell you it does not want to be stroked by chewing off your fingers. The Texas laws regulating fur-bearing animals prohibit keeping raccoons as pets.

Whether in the home or in the wild, a raccoon that loses its fear of humans can become a major menace. Wildlife rehabilitation groups have difficulty reintroducing them into the wild once they have become accustomed to human care. When released, one female yearling who had outgrown her human parents soon found the nearest ranch. She raided trash cans and even tore through a screened back porch to get dog food.

For years park engineers have tried to design a raccoon-proof trash can. Big Bend National Park once placed a new design at the campgrounds near Santa Elena Canyon. The can had a five-pound lid with a six-inch-wide steel border welded around the edge. The border overlapped the sides so that the heavy lid had to be lifted straight up. Shortly before midnight, a crescendo awakened me, and in the beam of my light a raccoon nonchalantly sifted through the refuse.

It sauntered to the next can, stood in the handle conveniently located on the can's side, and lifted the lid straight up. The carefully engineered, guaranteed animal-proof trash receptacle was child's play, no challenge at all for the adroit raccoon.

Trash cans are like credit cards to campground and suburban raccoons. Unlike their country cousins, these freeloaders seldom have to work for their meals. They may never wander more than a half-mile from home. The home range for urban raccoons covers about 12 acres, while their country counterparts require 40 to 100 acres to meet life's needs. The size of a raccoon's home range varies according to its age and sex, season, quality of habitat, and population density.

For raccoons, a good denning site makes a home complete. They will use a variety of shelters but favor a hollow tree for their winter bedroom and spring nursery. In the absence of a suitable tree, raccoons will snuggle up in brush piles, abandoned buildings, caves, crevices, or another animal's old ground burrow. As subdivisions engulf the countrysides, expectant mothers sometimes may set up house in attics and garages.

The mother raccoon has from three to four babies in April or May. Their eyes open in 18 to 24 days, and by 4 to 6 weeks they begin walking. By 7 weeks, the frisky babies run and climb. The mother raccoon has all the family responsibilities: she stays with the babies during the day and forages for food at night. After 11 weeks the young travel with the female on her nightly rounds. The young stay with her until the autumn, when they disperse. Sometimes the juveniles den with their mother for the winter and disperse the next spring. They reach their full size the second year.

The family foraging trips play an important role in preparing the young raccoons for life. Their instinctive knowledge equips them well for survival, but their mother's example shows them how to cope with the everyday problems of finding food and avoiding dangerous situations. Once while sitting on the banks of a bayou in East Texas, I heard a low whining growl interrupt the solitude. A movement at the water's edge caught my eye. A raccoon nervously paced back and forth crying under her breath. Wondering why she did not run away, I looked around and discovered three babies hanging motionless on a nearby tree trunk. The mother's distress calls had alerted the naive juveniles that danger was near.

Raccoons communicate by body language, as well as by vocalizations. An agitated raccoon lashes its tail, and when threatened it bares its teeth, lays back its ears, stands on its hind feet, raises its shoulder hackles, and arches its back. Many a hound has learned to respect these not so subtle signals from a cornered raccoon.

The ingenuity of a hunted raccoon has inspired and frustrated hunters since they first unleashed "coon hounds" in the woods. The baying call of the hounds sends the raccoons scampering. The normally slow-walking creatures can run surprisingly fast for considerable distances, climb as well as a squirrel, and swim like a duck. A pursued raccoon will try to get the dogs off its trail by climbing a tree, running along the limbs, and jumping out some distance away. Another trick is to head for water. Even the largest dog meets its match if it dives in after a raccoon: the raccoon will climb on the dog's back and hold its head underwater. Many a hunter has lost his prize hound to a swimming raccoon.

Raccoon fur has been popular since Davy Crockett wore a coonskin cap. During World War II, raccoon coats became the rage, but in the fluctuating world of fashion and furs, they dropped from view in the 1950s and 1960s. A revival in fur demand, especially overseas, brought the value of a raccoon pelt from a low of $1 to the present $15 to $30. Trappers catch more than three million raccoons each year in the United States. Animal furs support a $16 million per year industry in Texas. Our pioneer ancestors had to kill animals for food and clothing, but today we have other alternatives. I prefer to shoot raccoons with my camera and enjoy the antics of these entertaining creatures in the wild.

Fossil records indicate that raccoons have called North America home for more than one million years. Despite the increased pressures from trappers, hunters, and urban sprawl, these adaptable creatures are as numerous now as in pioneer times. If you live near water and woods, you probably have one of these enduring critters as your neighbor.

The Skunk: The Animal with a Chemical Defense

Striped skunk, polecat
Mephitis mephitis

What animal in North America, though the size of a house cat, does not fear any creature, large or small? All animals, tame or wild, respect the skunk, and usually let it pass unmolested. Skunks pay little attention to bears, mountain lions, wolves, or even humans. The bravery of this independent creature results not from brute strength, claws, or cunning, but from its ability to irritate.

Skunks have a chemical spray guaranteed to cause an attitude adjustment in an attacker. A direct hit with the spray leaves an indelible impression on the memory of the unfortunate victim. The noxious odor discourages most predators, but no defense is perfect. I have known dogs that regularly harassed and sometimes killed skunks. Usually the painful experience and the lingering odor remind animals to avoid future encounters with this harmless-looking animal.

A skunk's spray is as potent as Mace, as blinding as tear gas, and as noxious as rotten eggs. Powerful muscles around the pair of scent glands on the skunk's posterior eject the sulfur-based spray as far as 15 feet. When threatened, a skunk stomps its front feet, raises its tail, arches its back, and even prances on its front feet. Any animal that has encountered an angry skunk understands the warning and backs off. To spray its scent, the skunk looks over its back, raises its tail, and aims at the aggressor. My dog once challenged a skunk under our house. The barking ended abruptly with a yelp, and we heard him hit every floor beam as he blindly raced away. Moments later we evacuated our house.

Skunks seldom miss adversaries within nine feet. They can direct the musky liquid over an area of 30 to 45 degrees behind them, and vary the spray from a fine mist to rain-size drops. A sprayed animal, with eyes burning, lungs bursting for fresh air, and stomach retching, turns tail and runs. The only consolation to humans is that the symptoms eventually diminish and disappear.

If sprayed, you can remove the musk odor from clothes by several washings with detergent and household ammonia. Deodorize pets by bathing them with tomato juice or, more effectively, with neutroleum-alpha, a compound available from hospital supply houses. Use it in mop water to remove the odor from garages and basements.

The contrasting black and white back of a skunk makes it the most easily recognized animal in North America, both by humans and by other animals. That is exactly the purpose of the bold markings. As dusk settles across the countryside, skunks emerge from their dens as part of nature's night shift. Most nocturnal animals seek safety under the cloak of darkness, but not the skunk. Its warning colors advertise its presence to all. The white stripes reflect the fading glow of twilight and the light of the moon like a bright beacon warning others to stay away. While the colors of many animals help them blend in with their surroundings, secrecy is unnecessary for skunks. They have chosen to advertise their presence openly, mind their own business, and take what they want.

Because of their potent chemical defense, skunks have a dauntless attitude and aggressively pursue their every want. A naturalist at Big Bend National Park related that once, while he sat in a photo blind, a skunk ventured by. The hungry animal smelled the naturalist's fried chicken and invited itself for dinner. The unexpected guest received the greatest of courtesy. A skunk once interrupted my camp breakfast and devoured a carton of eggs mistakenly left on the ground.

Skunks waddle through the woods rooting for insects, spiders, small mammals and reptiles, fruit, nuts, and carrion. With short legs and a fat body, skunks avoid a chase or even a long walk. Lacking speed to pursue fast-moving animals, they lie in wait and slowly stalk mice, moles, ground squirrels, and other scampering animals. With a diet so easily satisfied, these animals seldom travel further than 1.5 miles from their dens. Rooting by starlight requires neither sharp vision nor a keen sense of hearing. Skunks depend on smell, their only well-developed sense, to find the insects and other small bugs that comprise 90 percent of their diet.

A skunk's smelly defense is not unique in the animal kingdom. Many beetles, one of its favorite foods, discharge foul-tasting secretions. Undaunted in pursuit of a tasty morsel, a skunk rolls the beetle in the dirt until it has exhausted its supply of chemicals. Just as skunks ignore the beetle's smelly protection, the great horned owl ignores the skunk's potent spray. Like most birds, owls cannot smell, and an airborne surprise attack gives a skunk little chance to establish its defensive position. Only owls and other large birds of prey regularly eat skunks.

As May approaches, female skunks begin searching for a protected den for their nursery. Though skunks root and dig with the five long claws on each forepaw, they prefer to use a natural or deserted den rather than digging one. The mother-to-be makes a nest of grass under a rock, in a thicket, in a dense stand of cacti, or in an old fox or badger den. By mid-May, from three to seven blind and helpless kits snuggle up in the safety of the den. After two to four weeks, the young have opened their eyes and are armed for action.

After two months, the kits accompany their mother on her nightly excursions. At first they follow single file and remain close to the parent, but quickly become adventuresome and wander further away. The young skunks have a short-lived home life, and by the end of August are rooting on their own.

Baby skunks look like adorable striped kittens, and at one time, de-scented kits were a popular item at pet stores. But skunks never quite become tame, and owners find they have to contend with a cantankerous, unpredictable wild animal in their home. The state regulations for fur-bearing animals now strictly control the killing and commerce of skunks. Keeping a live skunk requires a special authorization.

As a skunk matures, it becomes strictly antisocial. It never develops the manners and inhibitions of gregarious species, like dogs, that must learn how to get along with others. Except for mating and in communal dens, the adults do not interact. In Texas, skunks generally remain active all year, but in northern climates as many as ten may den together for the winter. At any time, several skunks may share the same sleeping den.

Skunks live in a variety of habitats, including urban areas. Their nocturnal habits and omnivorous diet enable them to coexist with humans. I have seen striped skunks on the University of Texas campus in Austin, and one has lived under my house in San Marcos, along with an opossum and my cat, for the past two years.

Suburban skunks sometimes cause problems. Besides occasionally spraying a dog or human, they may transmit dangerous diseases. They carry rabies more than any other mammal, with about 500 rabid skunks reported in the state each year. Outbreaks of the disease sometimes occur with a high percentage of skunks in an area becoming infected. Sick skunks may bite dogs or other animals and spread the deadly virus. A high percentage of skunks also carry leptospirosis, a bacterial disease that causes fever, muscle pain, and jaundice, as well as parasites harmful to humans.

In the wild, skunks are more beneficial than harmful. They eat prodigious amounts of insects harmful to range and agriculture, such as army worms, cutworms, bud worms, potato beetles, grasshoppers, squash bugs, and sphinx moths. They also eat large numbers of mice and rats, especially around farms. The fur industry considers the thick, durable, glossy fur of the striped skunk a favorite. Trappers catch thousands each year, especially in the colder climates that produce thicker pelts than Texas.

Striped skunks live in every Texas county, and range from northern Mexico to Canada. The twin white stripes from head to tail distinguish them from the other five species in the state. The smaller eastern spotted skunk, *Spilogale putorius,* and western spotted skunk, *S. gracilis,* have six broken stripes on their backs and sides. The eastern species lives east of a line approximately from Amarillo through Abilene to Austin and south to Laredo. The western spotted skunk lives west of a line approximately from Big Spring through Abilene, San Antonio, and McAllen.

The hooded skunk, *Mephitis macroura,* with markings similar to the striped skunk, has a distinct ruff, or hood, of longer hair on its upper neck. It occurs in the Big Bend region of the state and into Mexico. The hog-nosed skunk, *Conepatus mesoleucus,* has a single broad white stripe down its back. Its triangular range in Texas extends south of a line from Midland to Beaumont and back to Laredo, and into Mexico. The Gulf Coast hog-nosed skunk, *C. leuconotus,* lives along the coast from Aransas Pass south to Brownsville and into Mexico.

The Mysterious Jackalope

An animal so bizarre lives in the barrens of West Texas that most people deny its very existence. Only a lucky few have witnessed a jackalope bounding across the rugged West Texas plains, and many, like those who have seen UFOs, only reluctantly discuss their experience. Most people, especially in the scientific community, contend that such a strange creature cannot exist. But through the years enough evidence has accumulated to warrant a serious investigation of the facts. Unlike UFOs, jackalope sightings appear to have a regularity, both seasonally and diurnally. Sightings of these fantastic critters continue to increase as more people discover the salubrious climate of forsaken West Texas and retire to such sagebrush resorts as Terlingua, Lajitas, and Study Butte. Before I present previously unrecorded evidence supporting the existence of this alleged native unique to the desolate wastes of the Lone Star State, let me describe its physical appearance.

Fortunately we know exactly what jackalopes look like, which is more than you can say about Nessie the Loch Ness Monster, the Abominable Snowman, or Big Foot. The remains of jackalopes have been preserved by taxidermy, and are proudly displayed in homes and commercial establishments across Texas. Except for the trophy-size antlers crowning their heads, jackalopes bear an amazing resemblance to ordinary jackrabbits. Males apparently commonly display 10- and even 12-point racks. Unfortunately, to the chagrin of wildlife biologists, live specimens invariably elude all attempts at capture.

The antlers naturally make a jackalope immediately recognizable. If they wore such sporty headgear all year, they would probably be considered as common as sandburs. From an intensive study of all available accounts, written and oral, I have concluded that jackalopes, and just the males at that, have antlers only during the mating season, which lasts for a brief two weeks. During the rest of the year, the male's appearance is identical with the jackrabbit, and the female is indistinguishable at all times. Because of this, scientific studies have never determined the true population density, habitat requirements, demography, and social behavior of this chameleonlike species.

Many scientific studies have documented mimicry of one species by another. After all, butterflies copy each other's looks as much as the English copy the Queen's. The viceroy butterfly discovered that it paid to look royal, and copied the monarch's colorful pattern. Birds that eat monarchs get a stomachache that Alka-Seltzer can't touch, so the viceroy hides safely concealed in its royal garb. Once the reader fully understands the adaptive significance of mimicry, the advantages of the jackalope's deception are readily appreciated.

While photographs of Big Foot and the other mystery creatures invariably show blurred images of poorly exposed subjects disappearing in the distance, photographs of the jackalope show every detail in stunning clarity. The pictures render in sharp focus every hair on its sleek body and every bump on its rugged antlers. Jackalope photographs aren't hidden away in some dusty archive, either. Look for them at truck stops, souvenir shops, and cafes from El Paso in West Texas, to Orange, 855 miles to the east. Wherever postcards are sold, you can purchase jackalope photos for three for $1.

As I mentioned earlier, jackalope sightings occur regularly in the vicinity of Terlingua. The recent rebirth of Terlingua and the surrounding area is the fulfillment of a dream of the famous Union soldier General Philip Sheridan. He once said that if he owned Texas and Hell, he would rent out

Texas and live in Hell. Texas entrepreneurs have gotten rich hawking parched hillside ranchettes in the hottest spot in the nation to retirees trying to escape the arctic winters up north.

Another boon for Terlingua and jackalope documentation is the World Chili Cook-off held each year in the ghost town. Thousands of chili gourmets converge there, reminiscent of days of forgotten glory when the town boomed with frenetic activity. Once again music plays long into the night and more beer flows than water in the nearby Rio Grande. Most jackalope sightings occur after midnight, usually by a chili cook trying to find his camper.

That type of sighting corresponds with the second most frequent sighting of this elusive animal: near roadside taverns shortly after the 2 a.m. closing hour. Numerous studies have attempted to correlate jackalopes with roadside habitats; however, no statistically significant trend could be established, other than that they are nocturnal creatures. Recent Terlingua sightings completely discredit the roadside theory and reveal the startling details of the highly speculated, but never before reported, sex life of the jackalope. Once the animal's sexual behavior was understood, its entire life history fell neatly into place.

As previously mentioned, the male jackalope grows antlers only during its two-week rut, or breeding season. While in rut, the behavior of the normally docile, gregarious herbivore radically changes from Dr. Jekyll to Mr. Hyde. The males become bitterly aggressive, and vicious battles between rivals echo through the night. With antlers forward, the males charge each other at speeds exceeding 55 mph. About five feet before impact, they leap high in the air propelled by their powerful hind legs. Some accounts report that the midair collision is so violent that sparks ricochet from the antlers. The competing males repeat this charge-leap-collision pattern until the weaker is thrown backward with such force that its brain is addled. The bewildered animal loses his antlers and thinks he is a jackrabbit for the rest of his life. This behavior effectively removes the weaker animals from the breeding population.

The schizophrenia overcoming the male jackalope during rut affects more than his personality. The diet of this incredible animal undergoes phenomenal changes. During the brief period when they have antlers, the males become vicious predators and eat only meat. Their favorite food is coyote hearts, and they hunt the cunning canines with a passion. Terrified coyotes tuck tail and run, but they have no chance of escaping the speed demons chasing them. Is it revenge, or an increased caloric need for high-energy meat that drives jackalopes to pursue coyotes relentlessly? Only further research will tell.

The coyotes near Terlingua have gained a reprieve from the meat-crazed jackalopes. The enticing smell of exotic chili wafting across the desert distracts the ravenous jackalopes. They disdain coyote meat in favor of the gourmet concoctions that made Wick Fowler's name a household word. Only when observers realized that the roasting chili attracted the jackalopes were the hidden secrets of these mysterious animals finally discovered.

The territorial dispute between male jackalopes neatly answers a question that has puzzled children for centuries: how can Santa's reindeer fly? From my first memory, I couldn't enjoy Santa Claus because I knew deer couldn't really pull his sleigh through the air. But jackalopes leap with such powerful bounds that they could easily carry Santa's sleigh from house to house. When the truth is known, everything has a simple explanation.

If you ever have the mind-numbing experience of spending a day driving across the unfrequented deserts of West Texas, study the monotonous landscape for unusual-looking critters. Look closely at every jackrabbit you see, especially any that chase coyotes. For as sure as Big Foot lives in the California mountains, Nessie swims Loch Ness, and the Abominable Snowman treks across the Himalayas, jackalopes bound across the desolate West Texas plains.

Birds

Few groups of animals can match the diversity, flamboyance, and distribution found among the 8600 species of birds. They come in a rainbow of colors and vary in size from the two-inch, $^1/_{10}$-ounce bee hummingbird of Cuba to the 8-foot, 300-pound ostrich. In North America, the 5-foot whooping crane stands the tallest, and the 25-pound trumpeter swan and Canada goose weigh the most. The wandering albatross has the longest wingspan, 11.5 feet, of any living bird, while an extinct relative soared on wings stretching 20 feet from tip to tip. Birds exhibit an amazing diverseness even in flight. Some swifts, proving their name, can fly 200 mph, and acrobatic hummingbirds can hover, fly backward, and turn somersaults. Though strong fliers, ducks spend most of their lives floating and diving in the water, and roadrunners prefer dashing through thick vegetation chasing lizards. Forsaking the air, some birds have given up flight altogether. Penguins swim underwater using wings that have evolved into flippers. Other birds, like the kiwi and ostrich, prefer walking and running. Yet, despite the immense differences, all birds have one thing in common: feathers. Any animal, fossil or living, with feathers is by definition a bird.

Birds have mystified humans since the dawn of history. They appear in Stone Age art, Greek mythology, superstitions, folklore, and as ancient and modern religious symbols. Earthbound humans have considered birds messengers from the spirit world, representatives of God, and omens—for both good and evil. The popularity of birds continues unabated today, and they are perhaps the most loved of all wild animals. What accounts for this ageless love affair of humans with our feathered neighbors?

First of all, birds have the power of flight, an ability that humans coveted and eventually emulated at Kitty Hawk, North Carolina. Second, birds are exhibitionists. They flaunt their spectacular colors and remarkable singing talents at every opportunity. The brilliant pigmentation of a bird's feathers and the ringing melody it sings cannot fail to attract our attention and capture our imagination.

Scientists, trying to decipher the secret of flight from the feathered beasties, discovered that the ability to become airborne had not come without extreme specialization. Every feature of a bird's body has undergone remarkable modifications to adapt the animal for flight. Hollow and honeycombed bones replace the heavy, solid bones of the skeleton, and a lightweight bill substitutes for a mouthful of teeth. The flight muscles attach to an enlarged keel or breastbone, and bones in the arms, wrists, and fingers are fused and transformed into a wing. From head to tail, a bird's streamlined shape reduces drag, and the curve of the feathered wing creates the highest possible lifting force. Small feathers on the shoulder reduce the turbulence of air flowing over the wing, which would decrease the lift, and feathers on the wingtips act as tiny wings that enable the bird to fly at slower speeds without stalling. A bird's body, externally and internally, represents an engineering masterpiece of flight design.

Flight demands so much energy that birds must have special digestive, circulatory, and respiratory systems to supply food and oxygen to the wing muscles. Birds have the highest heart rate, body temperature, and metabolism of any warm-blooded animal. A bird's heart may pump between 600 and 1000 beats per minute to circulate blood, which contains twice the sugar content of human blood. Its stomach digests and assimilates food so rapidly that it must eat frequent meals throughout the day. Small birds daily may consume food equaling one-third their body weight, making nonsense of the expression "eating like a bird." Air sacs in their body cavity connect to the lungs and circulate air to increase oxygen absorption and to assist cooling.

Birds possess one of the most amazing features of any animals, a unique skin covering, feathers. Feathers, like fur on mammals, probably evolved not to enable flight but to conserve body heat for the warm-blooded bird. Fossils indicate that gliding, common today in many nonflying animals, preceded true flight. Feathers provide a waterproof coat and a highly insulating layer for warmth. To stay warm, a bird fluffs out its feathers to capture a thicker air space around its body, the same effect as putting a feather comforter on the bed. For their weight, feathers are the strongest, most durable objects found in nature.

Birds have two basic types of feathers. The pennae have a stout midrib lined with hundreds of tiny barbs. Barbules line the edges of each barb, allowing the bird to zip up the barbs much like a zipper. A bird can clean and repair a ruffled feather by passing it through its bill, an activity called preening. The flight feathers on a bird's wings, the tail feathers, and major outer body feathers are pennae. The down feathers, or plumulae, cover the bird's body and provide insulation. These fluffy feathers have only a short midrib and no barbules to hold the barbs together. The number of feathers on a bird's body varies with its size and the climate. A diminutive ruby-throated hummingbird has about 940 feathers, while 25,000 feathers cover a whistling swan. A bald eagle's skeleton weighs only 9.5 ounces, a fraction of its 24-ounce coat of feathers, illustrating the importance of feathers to a bird.

Besides providing a protective covering and enabling flight, feathers perform another vital function: they color the bird. Most birds have dappled brown, black, and white patterns for camouflage. Some, usually the males, sport brilliant colors designed to attract attention. The male's color and song warn other males of his species to stay out of his territory. The female, to increase her odds for rearing a successful family, chooses the male with the best territory. Some males have special feathers, like the peacock's flamboyant tail, for sexual display. Once the breeding season passes, the males usually molt and lose the nonfunctional feathers that decrease their ability to fly.

Flight has given birds a mobility unequaled by any other animal. On an ounce-by-ounce comparison, human flight and navigational technology cannot match the accuracy and flight capabilities of birds. Without map or gyroscope, many birds migrate thousands of miles twice a year. The whooping crane flies 2600 miles from Canada to the Texas coast, while the arctic tern flies 11,000 miles between its summer and winter homes with unerring accuracy. The ruby-throated hummingbird, weighing less than an ounce, flies 500 miles nonstop across the Gulf of Mexico.

The mobility of birds has allowed them to populate the earth and diversify like no other vertebrate. Birds have colonized every island and inhabit every continent, including Antarctica. They have adopted every imaginable lifestyle: birds eat seeds, sip nectar, catch flying, crawling, and burrowing insects, dive for fish, scavenge carrion, and kill other animals. These highly intelligent creatures even use tools, and some communicate with complex vocabularies: crows have 23 distinct vocalizations. Crows and blue jays sometimes use twigs to retrieve food just out of their reach. The woodpecker finch on the Galapagos Islands carefully prepares a stick to extract grubs from decaying trees. The dexterity of their bills and feet makes birds master architects in the animal kingdom. To protect their delicate eggs, birds weave intricate nests, usually single-family dwellings but sometimes multifamily condominiums.

Like a detailed résumé, the beak and feet of a bird reveal how it makes its living and where it lives. Seed-eating birds, such as sparrows and cardinals, have short, thick bills to crush the hard seeds, while insectivorous birds, like warblers and wrens, have slender bills to probe under bark and leaves. A woodpecker's skull and bill are designed to jackhammer holes in limbs, and its feet allow it to hang vertically on the sides of trees. The purselike bill of a pelican enables it to net schools of fish, and the webbed feet of waterfowl allow them to paddle and swim. The bill and long legs of a wading bird permit it to stalk the shallows for fish. Raptors have powerful talons to kill prey and hooked beaks to tear flesh. The vast diversity of avian beaks and feet illustrates the remarkable ability of birds to adapt to a multitude of environmental conditions.

Today approximately 8600 species and 28,500 subspecies of birds inhabit the earth. About 650 occur in North America, and 545 in Texas. Paleontologists can trace most of the modern families of birds back 11 million years to the Miocene epoch. Fossils of birds related to herons, flamingos, kingfishers, geese, and several others date back 60 million years. The oldest fossil showing feathers was found in Germany in 150-million-year-old limestone. The crow-size bird, named *Archaeopteryx,* had a small keel, indicating that the wing muscles probably were strong enough for gliding but not for sustained flight.

Fossils reveal that *Archaeopteryx,* which has no living descendants today, retained many features of its reptilian ancestors, who may have been some type of tree lizards. It had teeth, a long tail, and claws on its wingtips. *Archaeopteryx* lived during the same time period as the giant pterosaurs that dominated the skies but was not related to them. Though insignificant in comparison to the flying reptiles, birds survived the ages, while only the stones record the evidence of the once great pterosaurs. Fossils of the largest flying reptile yet discovered were found in Big Bend National Park. The creature, called the Texas pterodactyl, had a 35-to-40-foot wingspan and may have weighed 135 to 180 pounds. Scientists recovered about 200 bone fragments of the right wing, but none from the body.

Like most animals in the world, birds have suffered since humans began crossing the seven seas. About half of the more than 75 species of birds that have become extinct since the 1600s disappeared because of human hunting and habitat destruction. About 2 percent of the living bird species are in danger of following the dodo, passenger pigeon, and Carolina parakeet into oblivion. Observers have recent sightings of only one or two individuals of more than 80 rare species. Texas, with a diversity of habitats ranging from desert to seashore, mountains to prairies, and forest to brushlands, harbors more birds than any other state, including its share of rare, threatened, and endangered species. The Texas Parks and Wildlife Department currently considers 10 birds in the state as endangered and 18 as threatened with extinction. The following 11 sections describe the impressive capabilities, bizarre behavior, intriguing folklore, and human interactions of some of the most interesting birds in the Lone Star State.

Eagles: Masters of the Sky

Golden eagle
Aquila chrysaëtos
Bald eagle
Haliaeetus leucocephalus

In 1969, a voice rang out to all the world from 250,000 miles in space. Tension mounted as the Lunar Module used its last few seconds of fuel looking for a landing site on the cratered surface of the moon. Finally the dust settled and the triumphant astronauts radioed, "The *Eagle* has landed!"

What more appropriate name for the first spaceship to land on that distant moonscape than the bird that soars in the rooftops of the sky? From the time that ancient Persians and Romans carried eagles into battle, these majestic birds have symbolized courage, strength, and bravery. In 1782, Congress chose the bald eagle as the national emblem, despite an outcry that it had despicable habits. Bald eagles are infamous for forcing an osprey to drop its fish, and then catching it in midair. This thievery enraged early colonists who wanted a more respectable bird for the national emblem. But the bald eagle won the vote, and today it perches on the back of our one-dollar bills and adorns the Great Seal of the United States.

As aerial hunters, eagles are the undisputed masters of the sky. Nothing escapes their attention, even when they are soaring thousands of feet high. Their eyes see detail with five to six times the resolution of a human's. A human can discern a one-inch object at 30 yards, but an eagle can see it clearly from 200 yards. Such visual acuity enables these birds to spot a rabbit from almost 2 miles away.

The wings of such mighty birds are an engineering marvel for efficient gliding. Eagles can spread the tapering feathers on their wingtips to form many tiny winglets that reduce turbulence, increase lift, and prevent stalling at low speeds. A small feather, the alula, on the leading edge of their wing forces the wind smoothly over the top surface and prevents eddy currents from forming.

With a grasp far stronger than a human hand, an eagle's talons have legendary power. An eagle uses its powerful hind claw to kill small prey instantly while the three forward claws hold it securely. When a baby golden eagle sank its talons into a Montana scientist's arm while he banded the bird, his partner had to use pliers to loosen the eaglet's crushing grip. Similarly, an injured golden eagle in the care of a Texas wildlife rehabilitation group pierced its keeper's heavy hiking boot like an ax splintering kindling. The eagle's powerful bill complements its deadly talons; the upper beak has a four-inch hook to tear the flesh of its kill.

An adult eagle requires only about 8½ ounces of food per day, and it often gorges at one feeding and fasts for the next several days. These 9-to-12-pound, three-foot birds use different hunting strategies. The golden eagle hunts while soaring high above the ground. With a seven-foot wingspan, it glides on the rising thermals for hours without flapping, looking for its primary food, rabbits. A nest study in the Panhandle by the Texas Parks and Wildlife Department found 92 percent of the golden eagle nests contained the remains of black-tailed jackrabbits, 25 percent contained prairie dogs, and 21 percent contained the cracked shells of ornate box turtles. True to legend, eagles do drop turtles from great heights to break open their shells.

The bald eagle employs a perch-and-pounce method of hunting. It surveys its domain from high in a tree. When it spots a fish at the water's surface or a rabbit in the grass, it dives for the kill. Bald eagles usually have eaten sufficiently by 10 a.m., and then spend the rest of the day soaring effortlessly.

When Anglos first settled in Texas, the bald eagle nested in the coastal marshes and along river systems as far north as the Panhandle. With a diet primarily of fish, it lives close to the ocean, rivers, and lakes. Golden eagles, the sacred war birds of the Comanches and other American Indians, once nested across much of West Texas. These birds prefer the mountain peaks of the Trans-Pecos and the rugged canyons that dissect the rolling plains of the Panhandle. This cosmopolitan eagle inhabits desolate areas in North America, Scandinavia, Europe, North Africa, and parts of Asia. Because of the golden eagle's worldwide citizenship, our founding fathers favored the bald eagle, which is restricted to North America, for the national emblem.

Our national bird and its relative have not fared well with the taming of the continent. Believing that bald eagles, not the fishing industry, were depleting the rivers of salmon, the territory of Alaska placed a bounty on the magnificent birds in 1917. Within ten years, hunters killed more than 41,000. Before the bounty was removed, 100,000 bald eagles had died. Not until 1940, and 1952 in Alaska, did the slaughter of our national symbol cease and the birds receive protection under federal law. Today about 1000 pairs of bald eagles nest in the contiguous 48 states. The total population in North America, and the world, numbers less than 100,000, with some estimates as low as 35,000.

Golden eagles have fared no better. By the middle of this century, sheep and goat ranching began increasing rapidly throughout the Southwest, especially in areas overgrazed by cattle. The eagles found their habitat overrun with lambs and kids, and ranchers determined to protect their defenseless livestock. An eagle can easily kill a young lamb, and to most stock raisers, the only good predator is a dead predator.

Using airplanes, Southwestern ranchers gunned down 20,000 eagles in 20 years. One Texas pilot bragged that he had killed 400 eagles. Federal law protected bald eagles, classified as an endangered species, in all states, but the golden eagle received protection only under migratory bird laws and no state laws. Finally, in 1962, Congress passed the Golden Eagle Protection Act, which extended federal protection to the bird.

Though expert hunters, eagles never pass up a free meal. These opportunistic raptors care not if their meat is fresh or rotten. The mighty hunters feed on a dead carcass just as common vultures do. The habit gains them an unwarranted bad reputation in sheep country. Ranchers lose about 10 percent of their lambs each year, averaging 5 percent to natural causes, 4 percent to coyotes, and 1 percent to other predators. The sight of an eagle at a carcass does not prove that the bird killed the animal. Certainly eagles occasionally kill lambs and kid goats, and ranchers can be expected to want to protect their herds. But eagles have little or no impact on the sheep and goat industry. During the winter of 1982–83, the U.S. Fish and Wildlife Department Animal Damage Control Division in Texas received only eight complaints against eagles, and could confirm no losses of stock to the raptors. Yet many ranchers still hate eagles and shoot them on sight.

Coping with hunting pressure is only half of the struggle that eagles face in our modern world. The widespread use of DDT and related pesticides that began after World War II decimated the bald eagles more than the hunters' bullets ever could. The fat-soluble poison washed into the rivers and became concentrated in aquatic organisms. The poisons accumulated in the bodies of fish-eating birds until they could no longer reproduce; the pesticides killed the embryos and caused thin-shelled eggs that easily broke. Finally laws were enacted that prohibited pesticides that did not break down into harmless compounds, but not before bald eagles disappeared from much of their native range. Only now are these noble birds beginning to make a comeback.

Eagle populations recover slowly from loss because the birds take about five years to reach breeding age, and 75 percent of the young die before maturity. A pair of nesting eagles requires an average of ten years to produce two mature birds to replace them. In 1980, Texas Parks and Wildlife found 95 golden eagle nests in the state and 15 bald eagle nests. Each year about 85 golden eagles and 8 bald eagles fledge. Texas does not have many breeding eagles, but fortunately the population seems to be stable.

Eagles mate for life, and they may use the same nest, or aerie, for years, adding to it each season. Golden eagles favor ledges on sheer cliff faces in their mountainous habitat, and may alternate from year to year between several nests within their territory. Bald eagles nest in treetops along undisturbed waterways. One bald eagle nest measured 10 feet across and 20 feet deep and weighed two tons. Eagles begin nesting in February and March, and the female incubates the eggs for 40 to 45 days. After 9 to 11 weeks, by mid-June in the Panhandle, the eaglets can fly. In 1980, golden eagle nests in the Panhandle produced 1 fledgling per nest, and in the Trans-Pecos the birds averaged 0.8 young per nest.

Eagles usually lay two eggs, but the babies have a peculiar Cain and Abel relationship. The female eagle begins incubating the first egg as soon she lays it. She lays the second egg two to four days later, so it hatches two to four days after the first. The 3.5-ounce eaglets grow and develop rapidly each day. The four-day-old chick is already active and screaming for food when its helpless nestmate breaks through its egg shell. Even when the parents bring enough food to the nest for both, the older chick often pecks the younger to death. Puzzled scientists have never adequately explained the advantages of this homicidal behavior.

A growing threat to eagles, and all predators, is the inevitable urbanization of our countryside. Texas cities gobble up about 200,000 acres of open space each year. To survive, eagles need large expanses of rugged country. A nesting pair requires from 20 to 60 square miles of hunting territory for use exclusive of other eagles. Timbering, clearing for agriculture, flooding for lakes, and real estate developments continue to preempt large amounts of habitat from eagles and other wildlife.

As winter approaches, eagles from across the continent head south for warmer climates and more abundant food. Texas hosts the largest winter concentration of golden eagles in North America. About half of the population wings its way to the Panhandle, Trans-Pecos, and Central Texas. Between 500 and 1000 bald eagles winter in the eastern two thirds of the state, primarily along the coast, lakes, and rivers. You can see bald eagles wintering on Lake Buchanan (18 in 1985) by taking the Vanishing Texas River Cruise from November through February.

A New World for the Cattle Egret

Cattle egret
Bubulcus ibis

How often was America discovered? The Vikings arrived about 1000 A.D., Columbus in 1492, and the cattle egret in 1877. The pioneering attempts of the Vikings failed, but Columbus and the cattle egrets soon established a firm foothold in the New World. Both the human and the feathered explorers had strayed off course when they stepped onto the American shore. Columbus had set sail for India, and the egrets that landed in South America were flying across Africa. Presumably, a storm or strong easterly trade winds blew the migrating flock 2000 miles across the Atlantic. Like the helmeted conquistadors before them, the egrets, with reddish plumes adorning their heads, immediately began their conquest of the Americas.

Colonization was not a new concept to these airborne explorers. They probably originally called Africa home, but their inherent wanderlust drove them north across the European continent as far as the British Isles. The travelers flapped across Southeast Asia and island hopped to claim Australia as egret territory. Before they could call themselves true world citizens, however, they had to conquer one final continent. After that first flock reached landfall in Suriname, America was theirs.

The cattle egret is the only bird that has become naturalized on every continent, except Antarctica, without the help of humans. Excluding songbirds, it is the only bird that apparently has expanded its range during this century of human progress. What qualities enabled these amazing birds to spread around the world while most other animals diminished or survived only in a limited range?

As wanderers, cattle egrets can find new places, or habitats, with food and safe nesting. Humans, by converting forestland into expansive pastures, industriously create vast new homelands for these insect-eating birds. Unlike other egrets and herons that wade for their meals, cattle egrets prefer to dine high and dry in these man-made grassy fields. While other wading birds require the secure seclusion of swamps and coastal marshes, cattle egrets boldly march in step with cattle, horses, pigs, water buffaloes, rhinoceros, elephants, and even tractors. Wading birds suffer from the widespread pollution of estuaries and rivers, but cattle egrets feast alongside cattle in protected pastures safe from harmful poisons.

To colonize the new areas they discover, cattle egrets must reproduce in large numbers. Courting begins in April when the two-foot bird dons a wardrobe of cinnamon feathers that crown its head and accent its snow-white breast. Its bill, legs, and eyes change from yellow to scarlet, and they remain so until the eggs are laid. Large colonies of egrets build their nests in trees and bushes near water. Both sexes incubate the two to five eggs and meticulously care for the young. One adult always stays with the nest and quickly returns if frightened away, thus protecting the eggs and young from predators. With usually abundant food, they can double their population every year. The youth are the most prone to wander, and before the first breeding season often strike out on their own, seeking new lands to conquer.

By 1942, the relentless march of the cattle egrets had carried them through Central America, the Caribbean islands, and into Florida. With a "Westward, ho!" these pioneers advanced along the Gulf Coast to California, and north to Canada. They now breed as far north as southern Ontario, but retreat to the Gulf and further south in the winter. One was sighted 200 miles off the coast of Newfoundland. Perhaps it was an envoy to Greenland, still virgin territory for cattle egrets.

These indomitable birds reached the borders of the Lone Star State in 1955 and produced the first native Texan cattle egrets three years later. Those early homesteaders began rearing their families near Galveston. In 1959, only 11 pairs nested in the state, but the visitors liked their new home and six years later Texas boasted more than 20,000 breeding pairs. The total population exceeded 100,000 by 1972. Within a decade after they appeared in Texas, the V-shaped pattern of a flock of egrets silhouetted against a colorful sunset was a common sight along the Gulf Coast.

Cattle egrets have a feeding territory that they defend from intruding neighbors. Unlike the territories of most animals, their territory moves. The snow-white birds prance alongside a grazing cow, dart between its legs spearing grasshoppers, and sit on their benefactor's back. As many as ten birds jointly claim all insects flushed by their hooved ally. They take their positions around a cow like miniature cowboys and follow the animal as it moves through the pasture.

Cattle egrets eat more grasshoppers and crickets than any other food. Range-damaging invertebrates comprise 99.8 percent of their diet, along with occasional lizards, frogs, worms, and spiders. The easily satisfied birds even dine in garbage dumps when the opportunity arises. Since they eat no fish, they do not compete with other egrets, herons, or wading birds. Unlike starlings, house sparrows, and other animals spread by humans, these beneficial immigrants have no harmful effects on native bird populations.

One of the main secrets of the success of the cattle egret, besides wandering long distances and being opportunistic feeders and excellent parents, is the tolerance to humans these intrepid birds have developed. Some of their largest nesting colonies occur within sight of major highways. One of the first breeding colonies occurred along the shoreline around the busy Baytown Tunnel under the Houston Ship Channel. Cattle egrets nonchalantly forage along roadsides just a few yards away from speeding traffic. They even adopt tractors as their associates and follow the four-wheeled vehicles as they would a four-legged bovine. I have often seen the birds on the Clear Lake City golf course dodging balls and following golfers through the rough. These prolific birds, unlike most other wildlife, find the environmental practices of humans to their advantage. By converting the planet's woodlands into grassy pastures, we are creating a paradise on earth for the adaptable cattle egret.

Hummingbirds: Nature's Fanciest Fliers

Ruby-throated hummingbird
Archilochus colubris
Black-chinned hummingbird
A. alexandri

In the ironic world of nature, smallest sometimes means strongest. The hummingbirds, the second largest family of birds, have the double distinction of being the smallest warm-blooded (homoiothermic) creatures and the strongest fliers. The 1/10-ounce black-chinned and ruby-throated hummers zip through the air with a wingbeat of 50 to 75 times per second. During a courtship flight, a ruby-throat may beat 200 times per second! The English colonists, amazed at the buzzing sound of the tiny bird's wings, coined the name "hummingbird." In Spanish-speaking countries, these New World residents are called *chupaflor,* "flower sucker," *picaflor*, "flower pecker," and in Brazil, *beija-flor,* "flower kisser." My daughter excitedly called them "honey birds" when she was young.

The ruby-throated and black-chinned hummingbirds are the most common hummers in Texas. The black-chin breeds in the dry plateaus and deserts of the western half of the state. The ruby-throat, preferring the moist woodlands of the eastern half of Texas, is the only hummer that nests east of the Mississippi River. Their ranges overlap and both sip nectar in Central Texas. Six other hummers breed in Texas, and five regularly migrate through the state.

Despite their small size, hummingbirds migrate long distances. In the spring and fall, rufous hummingbirds migrate 2000 miles between Mexico and Alaska. Ruby-throats migrate 500 miles non-stop across the Gulf of Mexico. Contrary to folktales, the tiny birds do not hitchhike on the backs of Canada geese. They increase their weight 40 to 50 percent in stored fat prior to migration, and average 25 mph during the 26-hour flight.

Many birds can fly faster than hummingbirds, but none can match their maneuverability. Hummers can hover, spin like a top, and fly backward, sideways, and even upside down. When they hover, they beat their wings back and forth horizontally in a figure-eight pattern. On the backstroke, they turn their wings completely over so the bottom is facing up. They remain stationary because their backstroke is as powerful as their forward stroke. Other birds have a one-way power stroke, but not hummers. They get twice as much lift and propulsion from each wing beat. Of all birds, hummingbirds have the largest flight muscles in proportion to their bodies, equaling 20 to 30 percent of their weight.

If not speeding from flower to flower, these unique birds try to dazzle each other with aerial acrobatics. The ruby-throat and black-chin males loop the loop trying to attract females. Like a giant pendulum, they race up and down in the air in front of a female sitting on a perch. Often the male orients his power dive so the sun shines directly on him as he zips by his lady love. With each pass, he raises his iridescent feathers and flashes their brilliant colors. The lady seems to pay scant attention to the display, but who are we to say what she is thinking? If another male intrudes, the performer madly chases away the interloper.

Despite the sparkling red colors, hummingbirds have no bright red pigment in their feathers, only rufous, cinnamon, brown, and purple-black. The ruby-throat's glimmering scarlet throat comes from the way microscopic platelets in the feathers reflect and bend the light. Like prisms, platelets of different thicknesses and air content produce the gorgeous colors. When in the shade, a ruby-throated hummer appears dull gray. Thus, the male is camouflaged except when he displays in the sun.

The females of these two species have drab colors with

only a touch of green on their backs. Plain colors help them sit concealed in their thimble-size nests, hidden from predators. These birds build the smallest nests and lay the smallest eggs. A half-inch hummer egg is 7000 times smaller than the six-to-eight-inch egg of the ostrich. Hummers reinforce their two-inch nest with spiderwebs, which make it elastic enough to expand as the young birds grow. They line the inside with soft plant down, and decorate the outside with lichens. The two pea-size eggs hatch in about 15 days. After 2½ weeks, the babies can fly, but the mother feeds them for another 2 weeks until they learn where to find food.

The female does all the work raising the family, while the male busies himself flashing his iridescent feathers and chasing other males out of his territory. The brilliant throat and head feathers sparkle in the sunlight like warning beacons. The pugnacious male also fights vigorously for sole possession of a sugar-water feeder: when one has established a feeding territory, no other hummers enter unchallenged.

Hummingbirds have an avid curiosity and fearlessly investigate new objects. This probably helps them find new sources of food. To sustain their powerful flight, hummers require regular meals of sugar-rich nectar. Extracting the high-energy food from flowers requires hard work and specialized tools. Hummingbirds have a long bill to reach into deep-throated flowers, and a tubelike tongue with a fringed tip to sip the high-calorie liquid. Some tropical flowers, depending exclusively on hummers for pollination, are shaped and colored to attract the birds.

The color red universally attracts North American hummingbirds. Perhaps they find more red flowers while migrating. But hummers are creatures of habit—if most of the flowers in an area are blue, blue attracts them more than red. Hummers must eat about half their body weight in sugar daily. To stay healthy, they also catch flying insects. Captive hummers soon die if fed only sugar water.

To supply energy to their powerful flight muscles, hummingbirds have the highest rate of metabolism of any warm-blooded creature. Even at rest, they have a metabolism 50 times higher than a human's rate. These minuscule dynamos burn about 1000 calories daily. If humans burned energy at the same rate, we would have to consume 320 pounds of potatoes or 130 pounds of bread daily.

A hummer burns energy so fast that its system must slow down at night to conserve energy. Otherwise it could use so much energy staying warm that it might starve before morning. If its night temperature falls below about 93 degrees F, it usually enters a torpid state called noctivation, similar to hibernation. Its body temperature drops to near the air temperature. When its temperature falls to 60 degrees F, it burns only 1/60 of its normal energy. North American hummers take about 10 to 15 minutes to warm up and can fly when their body temperature reaches 86 degrees F. Their normal temperature is around 108 degrees F, depending on the species.

Hovering in front of a flower places tremendous physiological demands on the tiny bird's heart. Even at rest, the organ must pump 615 times per minute to supply enough oxygen-rich blood. When active, the heart may beat as high as 1260 times per minute. A hummingbird's heart represents 20 percent of its body weight. No other bird has a heart that large in proportion to its weight.

Small animals usually have a short life span, and one with the high-intensity heart beat and metabolism of a hummingbird could not be expected to live long. Yet both free and captive hummers have lived more than ten years. Perhaps noctivating contributes to a long life. Their exceptional flight skills certainly help them avoid predators: hummers fearlessly harass owls, hawks, and other nest-threatening animals, darting in much closer than other birds dare.

Once a hummingbird finds a rich source of food, it visits it regularly. It can even remember favorite locations from season to season. One spring, a female black-chinned hummer appeared outside our kitchen window hovering expectantly at the exact location of the previous year's feeder. We hurriedly filled and hung the jar, and to our delight she soon returned and feasted.

These tiny creatures often demonstrate their intelligence and ability to recognize who feeds them. The following story sounds like a fairy tale, but was reported in a scientific treatise on hummingbirds. A California artist convalescing from a serious illness in a sanatorium hung a hummingbird feeder in the window by his bed. Soon a male rufous hummingbird, a species ranging from Texas westward, claimed the feeder for his exclusive use. The artist delighted in seeing the feisty bird daily sipping the sugar water and chasing away uninvited guests. When the man was able to venture outside in a wheelchair, the grateful bird zoomed around his head and hovered directly in front of his face. This friendly greeting became a daily ritual. When the artist eventually returned to his country home by car, eight miles away, the faithful bird somehow managed to follow.

The artist took daily walks to build up his strength, always accompanied by his little friend. The rufous would wait for the slower human, and show him interesting sights he would have otherwise missed. His observant tour guide pointed out baby skunks, a covey of quail, and even a rattlesnake just off the trail. At last the day came when the man was well enough to return to his home in the city. After a month's absence, he returned, and stepping out of his car was greeted immediately by his feathered friend.

Within their feeding territory, hummingbirds have a regular circuit of nectar-rich flowers. You can be included in the bird's daily rounds by hanging a feeder with four parts water to one part sugar. Better yet, plant the bird's favorite flowers. Trumpet vine, honeysuckle, sweet pea, lantana, nasturtium, rose of Sharon, delphinium, and scarlet sage supply hummingbird cuisine. If you supply the food, you can enjoy the antics of these delightful birds all summer.

The Purple Martin: America's Favorite Bird

Purple martin
Progne subis

In some parts of the United States, the groundhog rates as the star performer of the winter season. According to legend, if it sees its shadow in early February, the winter chill gets another curtain call. But while people in the North search for groundhog shadows, people in the South gaze skyward eagerly anticipating the arrival of the purple martins.

Purple martins are the first migrating birds to arrive in Texas. Though weighing less than two ounces, these members of the swallow family fly to the United States and Canada from their winter home in the Amazon Basin of Brazil. They typically arrive in Central Texas around February 10. In warmer years, they may arrive several weeks earlier, and cold weather usually delays them. While some stay to nest in Texas, others move farther north, reaching New England by mid-April.

An advance guard of older males arrives first. These scouts often return to the same birdhouses that they used the previous year. By the time the females and younger males arrive several weeks later, the scouts have chosen a suitable home and are ready to set up housekeeping. After a week or so of relaxing in their summer home, both sexes begin constructing their nest. Martins do not build fancy nests, just a crude assemblage of twigs, string, and other scraps. Sometimes the birds line the nest rim with mud.

Martins originally nested in tree cavities, but since urbanization and modern timbering practices have made these scarce, they have adapted to city life. Like many people, these feathered suburbanites exchanged country living for an apartment lifestyle. Their penthouse must meet their specifications, though. A nest box at least 15 feet high and away from nearby limbs and buildings gives them security from nest predators. Their front doors should be from 2 to 2¼ inches in diameter and 1½ inches off the floor. Martins have learned that invaders can get through larger holes. An 8-by-8-by-8-inch room with ventilation holes seems ideal. The gregarious birds like to flock together, and prefer houses with at least six compartments.

Particular about their neighbors, martins avoid houses with nesting starlings or house sparrows. These two imported species aggressively drive away many of our native hole-nesting birds, including martins, bluebirds, and wrens. I saw a single pair of house sparrows destroy every egg in a colony of about 20 cliff swallows, and nest in one of their covered mud nests. You must continuously monitor martin houses in the early spring and remove the nests of unwelcome intruders before they become established. Sparrows and starlings tend to fill the nest compartment with grass, while martins mainly use sticks. A telescoping pole and hinged doors simplify matters greatly. Keep the nest entrances plugged until about a week before the martins arrive (a tennis ball fits the hole perfectly).

Once a martin family moves in, the birds busily swoop around their home catching food. They flap vigorously while chasing flying insects, then glide in long graceful dives. The males have a metallic blue-black back and front, while the females have a duller back and whitish belly. Like other swallows, martins prefer drinking on the wing. These agile birds even bathe while flying by dunking their rear quarters as they zip over the water, then shaking like a dog as they fly off. When not practicing their aerial acrobatics, the adults perch on the house and nearby wires and sing with a gurgling call. The delightful twittering song sounds like rushing water. In the morning, the whole flock joins the chorus to greet the dawn.

In Texas, martins begin nest building by late March or early April and lay their four to six white eggs from mid- to late April. Occasionally birds delay nesting until midsummer. The birds incubate the eggs an average of 16 days. A colony generally has babies hatching from mid-May to early June with parents busily zipping in and out feeding the hungry youngsters. Before the young fledge, they sit on their front porch and noisily beg for bugs from their foraging parents. One intrepid birdwatcher counted how often a pair fed their young between dawn and dusk. During the 16-hour period, the industrious parents made 205 trips! Martins need calcium in their diet and will eagerly feed on broken eggshells left near their home.

The young learn to fly in June and July, and the parents take them on increasingly longer foraging trips until they become independent. As the young fledge and begin spending their days foraging with the adults, the birds spend less time around the birdhouse. In July, the family packs up and leaves for winter retreat in Brazil. Like their spring migration, the birds set a leisure pace during the winter trip, with no hard and fast timetable. Once the martins leave, their house should be thoroughly cleaned and the entrances closed for the winter.

Bird enthusiasts love purple martins for their social nature, their airborne antics around their apartment birdhouse, and their voracious appetite for mosquitoes and other flying insects. Flying consumes so much energy that one of these active birds may eat 2000 insects a day. The residents of an eight-compartment martin house will consume about 32,000 insects a day, which is bad news for pesky mosquitoes.

The Western Electric plant in Kansas City used purple martins to solve its mosquito problem. Insecticides could not be used at the plant because of the delicate electronic parts being assembled, so the company installed martin houses around the buildings. The birds ate mosquitoes all day while the employees worked in insect-free comfort, and production increased significantly.

Despite the favor and protection of humans, purple martins have plenty of adversaries. Their chief foes are the starlings and house sparrows that usurp their nesting sites. Even though these two invaders can displace them, martins courageously defend their home against crows, hawks, and other predators. Screaming their loud alarm call, they mob any threatening intruder. Rat snakes can climb wooden poles and eat the eggs and young. For this reason, always place a section of sheet metal around the pole several feet below the house. Mites and other parasites can infest the nests and kill the young. Painting and sealing the inside of the house so the mites have no place to hide alleviates this problem. Occasionally, a screech owl may raid a martin house and snatch the birds right out of their nest. Proper construction, placement, and management of the house eliminates most of these problems, and experienced purple martin landlords often report that 100 percent of the babies survive.

Purple martins adapt well to city life and have delighted humans for centuries. The American Indians erected gourd martin houses in their villages. The colonists followed suit and constructed elaborate apartment dwellings for the friendly birds. Welcoming the purple martins is a 148-year-old tradition in Greencastle, Pennsylvania. Martin houses have lined the town square and streets since 1840. Increased interest and modern technology have in part repaid our debt to these displaced birds. Today purple martins are the favorite birds for neighborhood birdhouses. Gaily painted wooden structures and aluminum birdhouses especially designed for these much-loved creatures have sprung up across the country. The purple martin truly rates as one of America's most popular and beneficial birds.

The Mockingbird: The State Bird

Northern mockingbird
Mimus polyglottos

Our founding fathers may have given us the Lone Star flag, the cowboy, and other symbols associated with Texas, but the women of the state gave us the mockingbird. In 1927, at the urging of the Texas Federation of Women's Clubs, the Legislature declared the mockingbird the official state bird. They could have chosen the golden eagle, which winters across the western half of the state, or the cardinal that flashes his flaming red colors and greets the dawn with his ringing "whet-cheer." Of all the eligible birds, the women chose the mockingbird, and for good reasons. Perhaps no other bird in Texas is as well known and easily recognized, not because of dazzling colors, but due to its swashbuckling personality.

Never has a more fearless, defiant, belligerent, aggressive bird flitted through the treetops than the mockingbird. Nothing intimidates these indomitable birds. They will land at your feet to snatch up a grasshopper, or dive-bomb a cat. A mockingbird fears nothing that walks, flies, or crawls. One observer saw a mocker attack a diamondback rattlesnake until it pecked out the snake's eyes.

Let me tell you about the testy behavior of the JSC Mockingbird. When I worked at the Johnson Space Center, the antics of a mockingbird regularly amused the people in our office. She had declared the hedges bordering the walks outside our window her private domain. The mocker revealed her true personality in the spring when she nested. Most people passed within feet of her nest without noticing the drab gray bird hidden in the shadows of the leafy branches. On one occasion, a man wearing a hat saw the nest and pulled the limbs back for a better view. He might as well have disturbed a wasp nest. Luckily his hat protected him from the furious assault, but from then on, the fiery bird peppered anyone that passed wearing a hat. People generally tried to ignore the first few swoops of the feathered assailant, but soon discovered she meant business. Executives, politicians, and even astronauts ducked and ran unglamorously from the relentless barrage.

In the winter, the mocker raged constant battle against the cedar waxwings that invaded her privacy. The hungry birds wished only to feast on the juicy yaupon berries, an important ingredient of the mocker's winter diet. As soon as a flock landed in one end of the hedge, our feathered mascot streaked in like a Minuteman missile. The flock would scatter and reassemble at the other end, only to be pursued by the tireless defender.

Mockingbirds have the same bellicose attitude toward the hordes of robins that invade their territory every winter. Roy Bedichek, in *Adventures with a Texas Naturalist,* recalls the mocker that kept his hackberry tree free of robins when every other tree in the block overflowed with the noisy birds. Against such odds, other birds might have given up, but not the mocker. It sat on the highest branch and dived on any invading bird.

Most birds, and people, learn to avoid the wrath of these antisocial birds. The resident mockingbirds at JSC had divided the campuslike grounds of the Space Center into a patchwork of home territories. They did not tolerate trespassing. Once a crow came slowly flapping over just above the rooftops. As soon as it entered a mocker's airspace, the smaller bird attacked. Despite the dive-bombing, the crow maintained its course and leisurely pace. Each mocker would drop back when the crow left its territory and let the next resident resume the battle.

Despite their antagonistic attitude, mockingbirds mate for life, or at least for many consecutive seasons. In the winter, each has its exclusive and mutually respected territory. Come spring, the pair rejoins to rear its family. The male sings and protects the territory while the female incubates the three to six eggs for about 13 days and broods the babies. The beautifully colored eggs vary from pale to greenish blue with brown spots and blotches. Both parents energetically feed and protect the young. After they leave the nest in 11 to 13 days, the chirping cry of the hungry juveniles echoes through the bushes. I once found a baby mocker in the street a block from my home. I placed the almost fully feathered baby in an old grackle nest and set it on a rabbit hutch in my backyard. Within an hour, an adult, presumably its parent, found the baby and began feeding it.

Though the pugnacious mannerisms of the mockingbird make it conspicuous, its claim to fame rests not with its behavior. More likely, the women of Texas presented it to the Legislature in recognition of its lovely voice and remarkable ability to sing. Of all the birds in the United States, mockingbirds are the prima donna songsters. Other birds may have prettier songs, depending on your opinion, but none sing as long and intensely as our state bird. The mocker's ringing melodies fill the air from early spring into late fall, from daybreak until late into the night, especially during a full moon.

Despite the numerous poetic verses giving human emotions to the mockingbird's serenades, mockers, and all birds, sing for a more utilitarian reason: to warn others of their species to stay out of their territory. A mocker's singathons match its intense personality. Whether in song, in defense of nest, or in pursuit of insects, it acts with authority and assertiveness.

The intricate lyrics of the mockingbird's song have thrilled the human ear since the Indians first settled North America. The birds may incorporate as many as 35 bird songs into their own, and every mocker has its personal collection. The screaming "kee-ahrr" of a red-tailed hawk, the piercing "kill-dee" of a killdeer, the whistling "peter-peter" of a titmouse, and the honking "yank-yank" of a nuthatch may belong to the mocker's repertoire.

The mockingbirds' scientific name, meaning "many-tongued mimic," reflects their remarkable talent. Mockingbirds do not mimic in the sense of trying to fool other birds. That would neutralize the purpose of their singing. Their song tells other mockingbirds, "I'm the boss mocker, and you better stay away!" They certainly don't want other mockingbirds to think that they are a titmouse, chickadee, or a helpless baby bird, a call frequently used.

Numerous scientific theories and folklore attempt to explain why mockingbirds borrow so liberally from the melodies of other birds. One theory couples the many hours per day that a mockingbird sings with its expansive selection of tunes. If the bird sang the same tune repeatedly for hours, the competing birds of its species would become bored and ignore the defender's warnings. So, of necessity, the mocker spices up its act by singing eight to ten different tunes per minute.

An Indian legend presents another explanation of why mockingbirds use other birds' songs. At one time, only the mockingbird could sing. None of the other birds could make a sound, not even a peep. Naturally the other birds were intensely jealous. Finally the Great Spirit came and asked the mocker to teach his song to all the birds. The proud bird refused. The Great Spirit then asked him if he would teach each bird one verse of his song. The mockingbird agreed and did so until each bird had a melody, but some of the birds forgot their tunes. That explains why the birds sing only a part of the mocker's song, and some just quack, squawk, and caw.

The mocker's renditions of other birds' songs are not usually good enough to fool a careful listener. Some naturalists even claim that the similarity to other birds' songs is totally coincidence, but I'll stand by the belief that mockers have the remarkable ability to remember and repeat sounds. One study reported a bird that changed its tune 137 times in ten minutes, singing 43 songs. Another documented a hand-raised mockingbird that could sing 42 songs of 24 birds in a single performance. It also imitated a squeaky washing machine and often answered the calls of wild birds with their own notes. It even recognized birds and called their songs when they flew by.

The domain of a mockingbird includes not only the trees and shrubs in its territory but also the turf. They keep their sharp yellow eyes trained on the ground looking for grasshoppers, crickets, beetles, caterpillars, and other delectable creatures of the grass. Like a hawk on a rabbit, they drop from their perch and grab their prey. Often the hunt involves wing flashing, a behavior that birdwatchers have never been able to explain adequately. The mockingbird lands on the ground, takes several steps, and slowly and mechanically opens and closes his wings. Some suggest the white wing patches may help scare up insects. Only the mockers know for sure, and they aren't telling. Through the year, insects average 48 percent of their diet, accounting for more in the summer. Wild berries and fruit fill out their menu and comprise 85 percent of their diet in the winter.

In the days before air conditioning required closed windows to seal out the summer heat, I can remember awakening deep in the night and hearing the echoing melodies of the mockingbird's song. Somewhere in the dark shadows cast by the pale moon, the troubadour sang his medley to anyone who would listen. Somehow, the fluid melodies in the stillness of the midnight hours stirred within me a feeling of kinship with the sleepless bird.

Mockingbirds fit in well with city life. The manicured lawns of the sprawling suburbs supply an abundance of the mockingbirds' favorite insects. Two or three yards provide ample territory for the bird to rear its family, so every block usually has several resident mockers. In the wild, these birds prefer open habitat with shrubs and trees for nesting and cover. They rarely occur in the treeless prairies of the Panhandle and the deep woodlands of East Texas. In the summer, they range into southern Canada, and live permanently from northern California, through South Dakota, to

New Jersey, and south into Mexico. In addition to Texas, the states of Arkansas, Florida, Mississippi, and Tennessee have chosen this popular singer as their state bird.

The Owl: Warrior of the Night

Great horned owl
Bubo virginianus
Eastern screech owl
Otus asio

Deep in the still of the night, the "hoohoo, hoo, hoo" of an owl echoes in our mind long after the sound disappears. The somber cry ringing through a moonless night fills us with wonder—or fear. Throughout history, owls have symbolized the evil presences that lurk in the shadows. In the minds of many, these mysterious birds represent doom and desolation, dire misfortune, death's dreadful messenger, evil, and mischief. But in the topsy-turvy world of superstitions, owls also denote wisdom, intelligence, and scholarship. Few other birds have had such an influence on the thoughts of humanity as these feathered warriors of the night.

Part of the owl's mystique is its costume. The animal's soft downy coat conceals weapons of death. Its powerful talons can dispatch a small animal in an instant. Riding on silent wings, owls deliver death without warning, snatch up their unsuspecting victims, and disappear without a trace. Only their eerie call betrays their presence. No wonder so many cultures consider these nocturnal creatures harbingers of ill fortune.

Owls represent the nighttime counterparts to the diurnal birds of prey. Different species of owls fill the same niches that hawks, falcons, and eagles do. The great horned owl is the only eagle owl in North America. Like the bald eagle, it uses the perch-and-pounce method of hunting, sitting high in a tree looking for any movement on the ground. Other owls search for prey while soaring like hawks. Screech owls represent the evening complement to small insectivorous falcons, such as the kestrel. As true masters of the night, the owls' shape, plumage, and senses show amazing adaptations for nocturnal hunting.

Catching scampering animals by starlight requires a combination of sharp vision and keen hearing. Owls have flat faces and forward-facing eyes to maximize binocular vision, necessary for capturing moving prey. The size of their eyes allows the lens to gather enough of the faintest light to trigger the cells in the retina. The size, number, and concentration of cells and the way the cells connect to the brain make an owl's eye 100 times more sensitive to light than a human's.

Nocturnal animals usually sacrifice the ability to perceive detail, or visual acuity, for the ability to see objects in low light, but not owls. With a tube-shaped eye that spreads the light over as many retinal cells as possible, owls can discern fine detail in dim light. The tube-shaped eye is highly efficient, but every design improvement has its trade-offs. Owls cannot rotate their eyes, and they have a narrower field of view than humans. To compensate for the limitations, these specialized birds can swivel their heads 270 degrees.

Owls can hunt on the darkest night. The highly developed auditory portion of their brain provides the acute hearing necessary to locate prey in low light. Some species hear well enough to catch animals they cannot see. Their flat, disk-shaped face channels the slightest sound to the large ear openings on the sides of their heads. Feathers around the ear slits scan for noise like a mammal's large ears. The right ear is higher on the head than the left, providing directional hearing, an aid in pinpointing the exact location of a noise. Owls can discern sounds reaching the ears with a 0.00003-second difference. Screech owls and other owls that prey on small critters are most sensitive to high frequencies, like leaves rustling and squeaks. The great horned owl, which hunts larger animals, is most sensitive to 1000 hertz, the same as the human ear.

Acute hearing would provide little help in detecting and tracking prey if owls were noisy fliers. The flapping of their own wings would drown out any noises made by their quarry. To baffle the noise, the owls' wing feathers have soft edges enabling the birds to fly in almost complete silence. Their large wings also minimize the amount of flapping required.

With a 4½-foot wingspread, the great horned owl is the largest owl in Texas. This eagle of the night darts adroitly through forest branches and soars over desert shrubs in pursuit of scampering prey. Not finicky eaters, the aerial hunters dine on porcupines, skunks, rabbits, squirrels, mice, birds, and reptiles. The adaptable great horned owl ranges from Alaska and the subarctic through most of the Western Hemisphere to the tip of South America. They live comfortably in any semi-open terrain, whether rugged mountains, arid desert, humid seashore, or rolling brushlands.

The fury of a great horned owl is legendary. Two of these flying tigers once attacked a naturalist inspecting their nest. One drove its claws into the man's side while the other lacerated his arm with four-inch gashes. The owl's talon penetrated a tendon and partially paralyzed the arm. In another instance, an unwary hunter went to retrieve a great horned owl he thought he had killed. The injured bird sank its talons in his wrist and punctured an artery. Sometime later, both were found still locked in a death grip.

The diminutive screech owl is the common urban owl of Texas. It ranges from Alaska and southern Canada into northeastern Mexico. Unlike their larger relative, which favors wilderness settings, screech owls often live in wooded parks and along greenbelts in our most populous cities. At night, they sally forth capturing insects in midair with a loud snap of their beaks. Besides flying and crawling insects, they eat mice, moles, birds, lizards, and almost any other small animal.

Though not as powerful as the 20-inch great horned owl, the 8-inch screech owl is just as fearless. The grip of its smaller talons can barely pierce a toughened human hand, but with claws as sharp as a cat's, a swooping screech owl can deliver a slashing wound. Screech owls do not hesitate to attack larger creatures. Late one moonlit night, I looked out my bedroom window just in time to see my howling dog reel from the blow of a screech owl. His midnight serenade interrupted, the yelping hound ran for protection underneath the house. A small owl once blinded one eye of a person trying to photograph its nest.

Though hunters by night, owls sit stoically by day, their benign appearance giving no clue to their fierce disposition. Their sedentary behavior helps conceal them from other birds, which universally hold a great enmity for the bird-eating raptors. Great horned owls often seek shelter in a densely vegetated treetop or deserted building. Screech owls commonly roost in a hollow tree. Most birds recognize the upright, rounded profile of a roosting owl and sound a loud and excited alarm. Calling every woodland bird, the smaller birds furiously harass the owl, carefully staying out of its reach. Finally the sleepy bird gives up and flies away to find a more secluded perch.

Many times I have traced a loud clamor in the treetops to a group of birds mobbing an owl. Crows have a particular antipathy for the great horned owl, while blue jays despise the smaller screech owl. Once I released a screech owl that had recovered from an injury. The owl flew into a tree, and within minutes was discovered. Blue jays, grackles, titmice, and chickadees swooped and screamed until the confused owl flew away. Birdwatchers make good use of a bird's boldness when pestering a screech owl. They either imitate the whinnylike call of the small owl or play a recording. Every bird within hearing will flock to the sound and bounce around in the bushes in clear view. At night, the call brings the resident screech owl swooping to investigate the intruder in its territory. Calling screech owls can have unexpected results. One night a friend, who does an excellent screech owl imitation, barely missed the attack of a great horned owl diving in to snare its smaller cousin for supper.

Both screech and great horned owls have feather tufts

on their heads that resemble ears. The tufts break up their outline while they perch and make them more difficult to recognize. Their mottled and barred coloration helps camouflage them while sitting in the shadows. Screech owls may be either of two color phases to help conceal them in vegetation. The gray phase occurs most commonly in deciduous and hardwood forests, and the reddish phase among conifers, which have a redder bark.

Despite their name, screech owls do not screech. Their territorial and dueting call is a long, descending whinny with tremolo, followed by an even-pitched trill. Like most Texans, Lone Star screech owls speak a dialect distinctly different from those in the eastern United States. The western screech owl, *Otus kennicottii,* which can be heard in Big Bend and westward, has a call that sounds more like rapping on wood. Screech owls begin calling in February and remain vocal through September. They often begin their courtship calls shortly after sunset, and pairs call to each other through the night as they hunt. The pairs break up in September, and the small owls live alone until January.

Great horned owls are the classic hoot owls. With a deep bass voice, they utter four to seven soft tremulous hoots that resonate over remarkably long distances. Once, while camping at Big Bend National Park, I heard the soft "hoo, hoohoo, hoo, hoo" of the large owl all through the night and at dawn when I arose. It sounded as though he were just outside my tent. I followed the hooting for about a quarter-mile before seeing the owl perched atop a yucca stalk. Like the screech owl, these owls call more in the late winter and spring.

Owls begin courting and nesting in January and February, several months before most birds get the urge. Great horned owls lay their two to five round white eggs in a bulky nest high in a tree, often using the old nest of a red-tailed hawk. The eggs hatch in 28 days, and the young can fly in 10 to 12 weeks. Screech owls nest in tree cavities, often old woodpecker holes. The male feeds the female during the 26 days she incubates the two to six eggs. Both feed the young for 5 or 6 weeks.

Nesting early helps owls find food for the young before the dense summer growth of vegetation can hide small animals. Growing owls have voracious appetites, and screech owls may make as many as 70 feeding trips per night bringing insects and small prey to the young. One naturalist counted 6 rabbits in a great horned owl nest, and another measured 18 pounds of food in a nest, including 1 rabbit, 5 birds, 2 eels, 4 catfish, 11 rats, and 1 muskrat.

Two other owls common in Texas are the barred owl, *Strix varia,* and the barn owl, *Tyto alba*. Barred owls, named after the markings on their breast, live in woodlands, river bottoms, and swamps of the eastern half of the state. Their hooting call sounds like the bird is saying "who cooks for you, who cooks for you-all." Barn owls have a distinct heart-shaped face and white underparts. They prefer prairies, marshes, and open terrain, often inhabiting old buildings around farms, as their name implies. They do not hoot, but have a wheezy cry. A total of about 11 species of owls live in Texas.

The Roadrunner: The Paisano, Symbol of the Southwest

Roadrunner, paisano
Geococcyx californianus

"He must surely be the most comical bird in America!" laughed Texas author J. Frank Dobie. "He will go through more antics and cut more didos in an hour than a parrot can be taught in a lifetime!" Americans, particularly in the Southwest, share Dobie's enthusiasm for this strange member of the cuckoo family, the roadrunner.

The roadrunner comes by its name and reputation honestly. It would rather run than fly. This sleek, streamlined bird darts through the low brush and cacti of the Southwest with remarkable speed and agility. But while traveling at top speed, it will abruptly stop, raise its colorful crest, and look around as though pursued. Then, with a sudden burst of speed, it resumes its race with its head low and its tail parallel to the ground. This characteristic profile of outstretched bill and extended tail has inspired painters, woodcarvers, sculptors, potters, and souvenir-makers from Texas to Hong Kong.

Hollywood also recognizes the entertainment value of the bird with such strange behavior and speedy locomotion. Every American youth laughs at the famous cartoon series featuring the roadrunner and the coyote. The cartoon depicts the bird as a crafty jester that invariably outwits the inane Wile E. Coyote. This childhood introduction to the roadrunner justly might be considered modern American folklore.

The roadrunner has played a prominent role in the folklore of the Southwest for centuries. Along the Mexican border, the bird is known as paisano, meaning fellow countryman, or fellow traveler. Residents use the term to address someone who has experienced many of the same pleasures and hardships in life. The vaqueros who chased longhorns through the thorny chaparral of South Texas easily identified with the dauntless bird. Like the cowboys, the paisano also raced through the almost impenetrable vegetation, but instead of chasing ornery cows, it pursued lizards, mice, snakes, and other edible delights. A bird that had to wrest its livelihood from the same rugged country was an inspiration, and certainly a paisano.

The thick, shrubby vegetation where roadrunners live influences their curious behavior more than some imagined witty nature. The best way to find a gourmet meal of insects and small animals is by cruising along at ground level. A bird soaring above the treetops would have little chance of seeing or capturing a small creature in the dense vegetation. About the only time a roadrunner uses its wings is to sail from tree limb to ground. If pursued, a brief flight takes it into the thick brush, where it takes off running.

The roadrunner depends on legwork to keep its belly full. Catching the inhabitants of the thorny brush and desert requires fast starts, fast stops, quick turns, and quick bursts of speed. Some lizards can run up to 10 mph, but the roadrunner can top 15 mph. The birds have reportedly been clocked racing along roadsides at speeds exceeding 25 mph.

Though Hollywood depicts the coyote as the archenemy of the roadrunner, folktales emphasize its notorious battles with rattlesnakes. In one often repeated story, a roadrunner comes upon a sleeping diamondback. The crafty bird carefully constructs a corral of prickly pear pads or cholla cactus stems around the unsuspecting snake. Once the roadrunner has cornered the rattler with no route of escape, it pounces on the hapless reptile for the coup de grace. Using its heavy, slightly curved bill, the bird crushes the snake's skull with staccato hammerlike blows.

Roadrunners most assuredly kill rattlesnakes, and their victims are occasionally found with a crushed head and surrounded by cacti. A snake thrashing around in a cactus patch while being pecked to death could easily clear a small area. And that area might resemble a miniature corral. Thus, a little bit of factual evidence and some creative conjecture gives birth to an enduring folktale.

A roadrunner usually kills a snake for food, not out of animosity. Sometimes the bird's eyes are bigger than its stomach, and it catches a snake longer than it can completely swallow. A roadrunner hopping around with the snake's tail dangling from its mouth only adds to its comic reputation.

The bird's unusual vocalizations also contribute to the rich folklore surrounding this feathered warrior of the chaparral. In the spring, the male sits atop a mesquite or cactus and utters a guttural "crut-crut-crut." He bows his head low as though in prayer and slowly rises skyward while calling. The bird's tracks in the sand even resemble a cross. People thought that such a religious bird must be a paisano. The male actually sings his strange song to advise other males to stay out of his territory, and to attract females.

While courting the female, the male behaves like a true gentleman. He always brings a gift to try to win the lady's attention. He struts around with a scorpion or other delicacy dangling invitingly from his beak. If he is successful in wooing the prospective mother of his children, the pair mates. As often as not, the male does not relinquish his gift, and after the encounter, he nonchalantly swallows his present.

Unlike those members of the cuckoo family who lay their eggs in other birds' nests, roadrunners are dutiful parents. Both the male and female share equally in rearing the family, and each feeds the young chicks frequently throughout the day. The adult seems to freeze for a full minute with its beak thrust deep into the baby's throat while it regurgitates the baby's meal. When the young grow larger, the parents bring lizards and other small prey to the nest. Along the Texas border, roadrunners reportedly take parenting very seriously.

According to legend, roadrunners, not the stork, deliver human babies to their parents.

Despite their reputation for excellent parenting, roadrunners sometimes get confused. One spring, a female kept hopping up to the window ledge outside the ranger's station at Big Bend National Park. Behind the glass barrier, rangers had made a display featuring a roadrunner's nest. Finally, in frustration with the perfect home so close yet just out of reach, the bird laid an egg on the narrow windowsill.

Even though the roadrunner is the emblem of the wide open spaces and associated with the western deserts, this adaptable bird is equally at home in the palmetto swamps of East Texas. The chaparral bird likes open areas with scrubby brush cover whether in deserts, prairies, or broken woodlands. The bird delights its human neighbors from northern California east to Kansas, southward through Arkansas and northwest Louisiana, along the Texas coast to central Mexico, and west to northern Baja California.

In the Lone Star State, roadrunners dart across highways from the Louisiana border to El Paso and from the Panhandle to Brownsville. The bird occurs most commonly in West, Central, and South Texas, where the cowboy and all those who live close to the land still consider it their close companion. As J. Frank Dobie said of the bird he loved so much, "We true paisanos of mankind include in our kinship the paisanos of birdkind."

Talking Turkey in Texas

Eastern turkey
Meleagris gallopavo silvestris
Rio Grande turkey
M. g. intermedia
Merriam's turkey
M. g. merriami

Americans may vote dogs and cats their favorite animals most of the time, but for one week the turkey gets the honors. The warty-faced, fan-tailed creature is the center of attention on the fourth Thursday of each November. During the week, pictures of denuded turkeys, usually in a prone position with their feet in the air, appear in all the newspapers, and even on TV. The culmination of all this attention is the ceremonial carving and consumption of the baked and basted animal. Participants in the feast hoard scraps remaining after the feeding frenzy for future reenactments of the ritual.

The pioneer founders of this culinary festival were so eager to celebrate Thanksgiving year-round that the commonplace turkey completely disappeared from most of its original range. What few survived the overhunting had to contend with the conversion of their forest homelands into plowed fields and pastures. By 1930, only about 20,000 of these magnificent birds remained in just 21 states. States imposed hunting restrictions, outlawed hunting during breeding season when the males were the most visible, and initiated restocking programs.

Unfortunately most of the stocking efforts released crosses between domestic and wild turkeys, since the tamer birds provided easier targets. The tameness that made a hybrid variety a good game bird doomed it to certain death in the wild. Typically only about 1 percent survived the first six months. In addition, they spread barnyard poultry diseases in the wild population. State biologists finally overcame the political pressure to use birds bred for easy hunting, and the restocking programs became successful. Shooting a turkey once again became a true test of a hunter's skill.

All of the three varieties of wild turkeys originally lived in the Lone Star State. Hunters gunned the Merriam's turkey out of the Guadalupe Mountains before the 1900s, and the last eastern wild turkey from the Panhandle in 1905. By 1942, less than 100 turkeys survived in the East Texas forests. The Rio Grande turkey ranges from South Texas through the Hill Country into North Texas and Oklahoma. It is the only variety that, though its numbers greatly declined, did not suffer a drastic reduction in range. Today biologists have restocked the Rio Grande turkey successfully in the Panhandle and through much of its native range. The gobble of the Merriam's turkey once again echoes through the Guadalupe Mountains, as well as the Davis Mountains. The future of the eastern turkey in East Texas remains less certain. Lumbering conglomerates own most of the pine-hardwood forests and annually convert hundreds of thousands of acres of the rich woodlands into barren pine plantations devoid of food for wildlife.

Besides their keen senses and a tendency to flee at the slightest alarm, wild turkeys survive because they eat almost anything. Their favorite foods include nuts, acorns, juniper berries, grass seeds, and insects, but they will not scorn an occasional lizard, snake, or snail. When they find a rich resource, they gorge themselves in true turkey fashion. By storing up fat, they can survive the harshest winters. They have been known to go 24 days without food and lose 40 percent of their weight, and still survive.

Turkeys form flocks of 30 to 40 in the fall after the poults, or chicks, have matured and the spring breeding aggressiveness has disappeared. The birds like to roost in large deciduous trees near water, and flocks sometimes return to the same site for 100 years. The 15-to-20-pound gobblers and 6-to-8-pound hens reach the highest limbs by short flights from branch to branch. They come to the roost as the sun sets and leave at dawn.

Turkeys have contributed to human life in North America since the Indians first domesticated them 2000 years ago. After almost disappearing from its ancestral range, the wild turkey population has increased to 2 million strong, with 500,000 in Texas.

[At this point in the chapter, I was overcome by a trance and began automatically writing under the direction of Ben Franklin. This may be the first time in recorded history that a spirit used a word processor to speak to posterity. I present the following unedited paragraphs to the reader. The facts about turkeys are accurate, but I have no way to ensure the authenticity of certain historical references—G.M.]

Even though every American history book tells how turkeys came to be the main attraction of Thanksgiving, the complete story has never been told. The choice resulted from a conspiracy, not by the bloody British but by the Puritans. After over 200 years of secrecy, I will now reveal the truth to you fortunate readers.

The colonists admired the turkey almost more than any other animal. The wild turkey reigned as the king of the game birds. With a sleek, streamlined body, powerful legs, and keen eyesight, it streaks through the woods like a racehorse, reaching speeds of 15 mph. If danger presses, it bursts into the air with a strong, swift flight, though it seldom stays airborne for over a quarter of a mile.

At a time before TVs or comic books, the colonial youth looked to male turkeys as a chief source of amusement. The gobbler's sexual behavior, a subject taboo in all forms of polite Puritan society, created the main interest, and provided ample subject matter for ribald jokes and stories.

In the spring, the male's wattle-covered face turns as red as Old Glory's stripes. He struts around with a bounding stride chasing away other toms and courting the females. Evidently the sight of the malshaped crimson face and long hairlike beard growing from his chest strikes admiration in a lady's heart. The tom advances toward a hen with a subdued gobble and his tail fanned while quivering his dropped wings. After the gobbler enjoys the company of one hen, he struts on looking for another. Since a male turkey has a large number of females in his herd, being called a turkey in colonial times was the epitome of a macho compliment.

The Puritan leaders, however, had a contrary opinion of the male turkey, an animal that seemed to live for one inexcusable purpose: sex. On top of that, the cursed animal deserts the female after getting her in the family way. The lady retreats to a secluded nesting spot and hatches between 10 and 14 eggs with the father assuming no responsibility for the children. Such behavior provided the topic for many a Sunday sermon. The Reverend Cotton Mather became famous for his Turkey Sermon.

Despite being slandered on Sundays, the wild turkey captured the imagination of the more daring colonists. Its peculiar gait inspired a dance, the turkey trot, as popular on Saturday night as a picnic on the Lord's day. Ben Franklin even led a movement to adopt the turkey as the national bird.

How could the Puritan leaders squelch the admiration for turkeys held by the country's youth and political leaders? Outright condemnation from the pulpit had failed, so they launched an indirect assault on the turkey's reputation. To save the impressionable youth from the influence of such an aberrant bird and to preserve the honor of this great nation, the T-Day conspiracy was born. The strategy was to make the turkey the center of the emerging culinary festival called Thanksgiving. Who could admire an animal eaten with relish, or even cranberry sauce, every year? The plan worked, and any reference to gobbling food or being called a turkey became an insult.

[Thanks, Ben, for the inside info.]

Vultures: Nature's Garbage Disposals

Turkey vulture
Cathartes aura
Black vulture
Coragyps atratus

"Ranger, we thought we had killed the poor bird! He flew up right in front of us and bounced onto our hood. He threw up all over the windshield before he flew away!" The ranger at Big Bend National Park tried not to laugh at the gentleman's encounter with the turkey vulture. Visitors to the park often have an unglamorous introduction to the ungainly bird. The ranger explained that vultures naturally regurgitate when threatened or frightened. They have a hard enough time getting airborne without a headwind, and the added weight of an engorged stomach doesn't help a quick take-off.

I had already discovered the vulture's disgusting habit of jettisoning its stomach contents. The Austin Nature Center had an injured turkey vulture named Groucho Marx, and my job was to make him (or her) feel at home. Groucho had a nasty habit of coating my shoes with goopy green stuff every time I entered his cage. I never knew whether he related to my shoes like a dog marking a tree, a skunk spraying a foe, or a mother vulture feeding her baby. To my relief, he finally decided to keep his supper to himself, and merely shredded my shoelaces.

One day, we tried to liberate Groucho. We took him to the nearby woods and cast him into the air, expecting him to wing his way to freedom. Instead he glided to the nearest picnic table and contentedly folded his wings. After several hours, we realized that Groucho wasn't about to trade a safe perch and a full food bowl for a hard day's work in the sky. I wasn't sure how to get him back to his cage. He didn't want to leave the table, but he couldn't resist my shoes. I led him, Pied Piper style, back to his cage.

Vultures have such a hard time finding food that they only reluctantly abandon a meal, whether in a cage or on a roadside. Safe roosting sites are also at a premium. No wonder Groucho had no ambitions to abandon human charity. In Big Bend, turkey vultures find an ideal roost in the tall trees shading the campgrounds at Rio Grande Village. The large cottonwoods bordering the river provide a safe haven for a large population. At dawn, they noisily flap their three-foot wings and glide from their roosts high in the trees to the closest picnic table. After defacing the table, they hop awkwardly around on the ground until some photographer harasses them into flying. The unsightly birds remain earthbound until the sun has warmed the desert enough to create thermal updrafts.

When vultures take to the airways, they become a picture of grace soaring on the rising thermals. They ride the updrafts like a kite circling higher and higher until they become mere dots. These perfect sailplanes can glide for hours and cover miles of terrain without expending the energy of a single wing flap.

Every feather on a vulture's wing is designed for one purpose: gliding. Feathers extending beyond the wingtips act as tiny, narrow wings and prevent stalling at low speeds. They also reduce vortices along the wing edges. The broad wings give the bird a high lift factor, essential for long-term soaring without flapping. The rounded design of the wing allows the bird to maneuver in the tight circles necessary to track rising columns of air. Even the vulture's size suits its lifestyle of riding the thermal updrafts. It is large enough to provide stability in the capricious currents, but light enough to require a minimum of energy to stay airborne.

While the red-headed turkey vulture and its black-faced cousin, the black vulture, soar through the skies, their eyes remain glued on the earth below. With keen vision they scan the countryside for food. Unlike birds that kill their food and eat it fresh, vultures feast on dead animals, and the riper the better. The weak talons of these birds of prey cannot grip tightly enough to hold a live animal or to carry objects very far, and their slightly hooked bill is too weak to kill. By grouping together for a mass attack, however, vultures can kill a small animal like a skunk.

Turkey vultures have an advantage over black vultures in finding food. Both rely on their sharp vision to spot carcasses in the open, but turkey vultures have an additional sense missing in almost all other birds: they can smell. Their brain has a well-developed olfactory lobe. Smell guides them to prey hidden from sight in brush or trees until they can locate it by sight. Despite their acute sense of smell, they must see the prey to find it.

Vultures willingly serve as natural garbage disposals; they are as important as trash collectors in a city. They congregate in large numbers around the carcass of a cow or deer. A half-dozen may sit on fence posts and in nearby trees while several feast on a road-killed armadillo. In Central American countries, black vultures often eat rotten fruit and vegetables. Both species gather in large numbers in garbage dumps. These normally silent birds hiss, grunt, and growl at each other while dining. In a mixed group, the smaller but more aggressive black vulture often drives away the turkey vulture.

Despite its filthy occupation, a vulture retains a tidy appearance and, even though it sometimes wallows in its food, never lets offal besmirch its feathers. Dirty feathers mean inefficient flying, a costly mistake for a soaring bird. A vulture cleans its feathers by preening, running them through its bill, but it cannot reach its neck and head feathers, the ones most likely to get soiled. For improved sanitation, vultures are endowed with a naked head and neck.

Vultures, or buzzards as people commonly call them, prefer open country where they can see dead animals from long distances. Turkey vultures, the most abundant vulture in Texas, occur throughout the state, but are usually rare in the Panhandle and Trans-Pecos in the winter. They range from southern Canada to the Tierra del Fuego at the southern tip of Argentina. Turkey vultures reside year-round through most of their range, but must migrate from areas where winter temperatures freeze carcasses. The birds soar 3500 miles to southern climates, fasting for the two-week trip, and fly about nine hours a day, averaging 40 mph.

Black vultures range from Maryland south through the coastal states to central Argentina. In Texas, they commonly live in the eastern and southern halves of the state. Black vultures, probably the most numerous birds of prey in the Western Hemisphere, occur most abundantly in the tropics. They thrive in a hot climate because they are not as efficient at soaring as the turkey vulture, and they require stronger updrafts to stay airborne. Black vultures have shorter wings, flap frequently, and use more energy while hunting, but the tropical thermals compensate for the inefficiency. The rate at which a soaring bird descends in calm air is a measure of its gliding efficiency. Black vultures sink at a rate of 0.79 meters per second, while turkey vultures sink at 0.61 meters per second. This slight difference translates into a considerable additional energy expenditure for the black vulture over a year's period.

In the last three decades, the number of both vultures in Texas has decreased drastically. Many factors have contributed to the decline. Less carrion is available. The blowfly, which infected and killed many cattle and deer, was brought under control by 1950. Ranchers deprive vultures of food by burning carcasses to prevent the spread of disease. With less carrion available in the fields, vultures must rely on road-killed animals for sustenance, and sometimes themselves become victims of speeding vehicles. Ranchers habitually kill vultures because they believe they transmit diseases, which has never been conclusively proved. In addition, habitat destruction has eliminated nesting sites and prey availability.

The black vulture has suffered much more than its red-headed relative. In a marginal temperate climate where competition is the keenest, the turkey vulture's advantages of a sense of smell and more efficient soaring give it the edge. The turkey vulture's increased mobility enables it to search over more ground for scarce food while using less energy. As the years pass, the black vulture slowly retreats south, where optimum conditions exist for its survival.

Soaring black and turkey vultures are easy to tell apart. Look for the turkey vulture's characteristic V-shaped profile. The bird gently rocks back and forth as it glides along. The entire edge of the wings from body to tips appears grayish white. The black vulture holds its wings flat as it glides, and only the wingtips are whitish. At close range, the turkey vulture's red head is readily distinguishable, though juveniles have dark heads.

The Armadillo, *Dasypus novemcinctus*

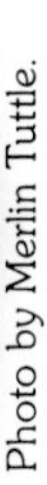
Photo by Merlin Tuttle.

Mexican free-tailed bat, *Tadarida brasiliensis*

Plate 1

Cougar, *Felis concolor*

Coyote, *Canis latrans*

Plate 2

White-tailed deer, *Odocoileus virginianus texanus*

Bottlenose dolphin, *Tursiops truncatus*

Gray fox, *Urocyon cinereoargenteus*

Javelina, *Dicotyles tajacu angulatus*

Plate 4

The Longhorn

Opossum, *Didelphis virginiana*

Pronghorn, *Antilocapra americana*

Black-tailed jackrabbit,
Lepus californicus

Jackalope
(no designated scientific name)

Raccoon, *Procyon lotor*

Hog-nosed skunk, *Conepatus mesoleucus*

Bald eagle, *Haliaeetus leucocephalus*

Golden eagle, *Aquila chrysaëtos*

Cattle egret, *Bubulcus ibis*

Ruby-throated hummingbird, *Archilochus colubris*

Plate 9

Purple martin, *Progne subis*

Mockingbird, *Mimus polyglottos*

Great horned owl, *Bubo virginianus*

Eastern screech owl, *Otus asio*

Roadrunner, *Geococcyx californianus*

Eastern turkey, *Meleagris gallopavo silvestris*

Plate 12

Turkey vulture, *Cathartes aura*

Black vulture, *Coragyps atratus*

Golden-cheeked warbler, *Dendroica chrysoparia*

Whooping crane, *Grus americana*

Plate 14

American alligator, *Alligator mississippiensis*

Southern copperhead, *Agkistrodon contortrix contortrix*

Monarch butterfly, *Danaus plexippus*

Red-legged tarantula, *Dugesiella hentzi*

Western diamondback rattlesnake, *Crotalus atrox*

Western cottonmouth, *Agkistrodon piscivorus leucostoma*

Texas horned lizard, *Phrynosoma cornutum*

Baby Kemp's ridley sea turtle, *Lepidochelys kempi*

Three-toed box turtle, *Terrapene carolina triunguis*

Imported red fire ant, *Solenopsis invicta*

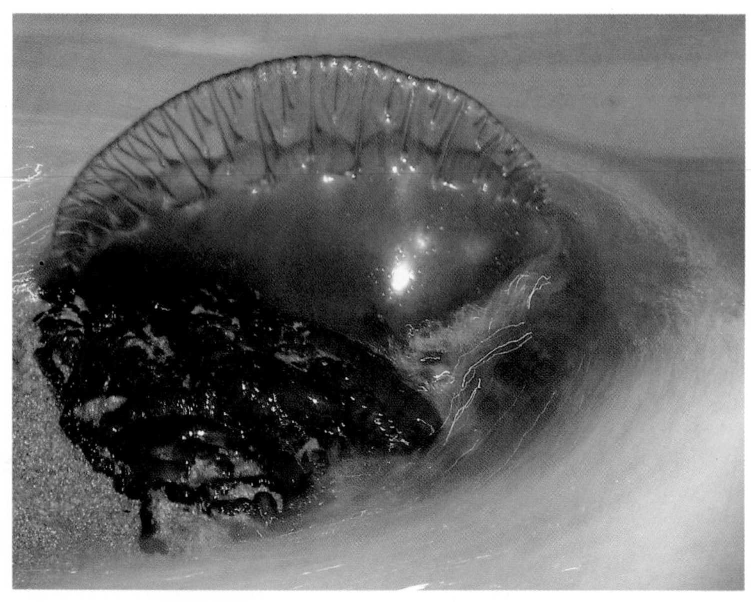

Portuguese man-of-war, *Physalia physalia*

Striped scorpion, *Centruroides vittatus*

The Golden-cheeked Warbler: A True Native Texan

Golden-cheeked warbler
Dendroica chrysoparia

Each spring, hordes of enthusiasts from around the nation descend on the cedar-covered hills of Central Texas searching for a treasure of gold. Instead of panning the many spring-fed streams for precious nuggets, they scan the treetops with high-powered binoculars. Their quest is the rare and endangered golden-cheeked warbler. After spending the winter in Guatemala, Honduras, and Nicaragua, these tiny birds migrate to the Texas Hill Country to rear their families. Every golden-cheeked in existence is a native Texan, a distinction held by no other species of bird.

What unique feature in the Hill Country attracts these colorful birds? The cedars, the same trees despised by sniffling and sneezing humans for spreading highly allergenic pollen, attract the golden-cheeks. Each winter, misery floats on every breeze as the trees shed copious amounts of the pernicious pollen. By spring the air has cleared, and the energetic golden-cheeks, like feathered butterflies, enliven the aromatic branches. The breeding range of the golden-cheeked warbler coincides with the distribution of the Ashe junipers, commonly called cedars, across the Edwards Plateau.

Excitement comes in small packages for birdwatchers. The 4¼-inch golden-cheek weighs only a third of an ounce. The birds typically arrive in the hills west of Austin around March 12. One year on that date, I met a birdwatcher from Ohio at the Travis County Audubon sanctuary. She had scheduled her birding trip to spend that one day in Austin hoping to add the rare warbler to her life list. Many birdwatchers keep a list of every species seen, and some keep it with religious zeal. The bird books gave the average arrival date of the golden-cheeks, and she was ready with binoculars in hand. The birds, however, operate on their own timetables, and to the woman's consternation, the warblers were tardy that year. Without forgiving the belated birds, she rushed off to her next stop, still with an unmarked square beside the elusive golden-cheek on her checklist.

I remember the excitement the first time I saw this colorful warbler. The cedar brakes were stifling, with April temperatures already in the mid-90s. I had heard several singing males, but hadn't found them. Suddenly one popped up in the tree above me and began singing. I caught my breath when I saw him through my seven-power binoculars. No picture can capture the intensity of his jet black head, throat, and back contrasted with his blazing golden cheeks. His buzzy song seemed to burst from his throat, then the ball of energy disappeared into the brush. The females, also with golden cheeks, have a dull olive back and streaked throat. Both sexes have white bellies.

The female golden-cheek builds her nest without any help from the man of the family, who stays busy singing. She adheres to a very strict building code established by thousands of years of tradition. The ladies demand the rustic look of cedar, and always construct their homes with long strips of cedar bark. They pull loose strands from the limbs and, using spiderwebs to hold them together, weave a three-inch cuplike nest. As a finishing touch, they line the interior with soft rootlets, mammal hair, and feathers.

The males are far from idle while the females build the nest, lay, and incubate the three or four eggs. In the division of labor, each sex has an equally important role. All day—from the time he first arrives from his southern retreat—the male sings his wheezy song to proclaim his territory. Any intruding male meets a furious assault. The males battle out a division of the turf, or more properly the treetops because they never forage on the ground. Each male stakes out a three-to-six-acre domain with enough resources to rear a family successfully. Females choose aggressive mates that will protect the nest, vigorously defend their territory, and help feed the young.

Once the eggs hatch, after 12 days of incubation, the male becomes a family man. The dutiful father sings less and less as he devotes more time to feeding the hungry babies. From dawn to dusk, both parents fill the gaping mouths of the young with caterpillars, spiders, flies, moths, and any other insects they can catch. After 9 days, the young leave the nest, but remain under the protection of their parents. They learn to forage on their own in about 3 weeks. The parents appear to divide the family, each taking care of the young in its charge.

Golden-cheek nesting coincides with the peak caterpillar population in the plateau live oak, Texas oak, and other hardwood trees. The birds scour the limbs gleaning the tiny caterpillars, the major food for the voracious appetites of the young. Without the insects in the hardwoods, the young would starve. Golden-cheeks never nest in pure stands of cedar, nor in cut-over second growths. Only mature cedars, at least 50 years old, have bark that easily shreds from the trunk and limbs, so only the association of mature cedars and hardwoods provides a suitable habitat for this rare bird.

Golden-cheeks nest in 31 counties in Central Texas, but exclusively in the remaining patches of mature cedars, which are becoming as rare as the bird so dependent on them. The cedar eradication program of the Soil Conservation Service, land-clearing for pastures and agriculture, and sprawling urbanization destroy the warbler homeland at an alarming rate. A 1974 field survey estimated that in the 150-by-50-mile wedge-shaped area occupied by the birds, only 326,000 acres of suitable habitat remained, and almost half of that was considered marginal. At that time, the world's population of golden-cheeks numbered about 15,000.

Habitat destruction may eventually eradicate the golden-cheek, but the birds also have their share of day-to-day worries. The babies often suffer ill-treatment from unwelcome guests. Besides having their home burglarized by rat snakes, golden-cheeks are victimized by the brown-headed cowbird, one of nature's cleverest con artists. In some areas, these birds parasitize as many as 60 percent of the golden-cheek nests. Cowbirds prefer to let someone else shoulder parental chores, and they lay their eggs in the nests of warblers and other smaller birds. The female cowbird closely watches the other bird's nest and lays her egg just before the host begins incubating. She discards several of the rightful eggs to make room for her own. The golden-cheek often recognizes the foreign egg and abandons the nest to start over. Once a cowbird hatches, the warbler babies are doomed because the aggressive baby kills the other young. The foster parents dutifully feed the cowbird until it leaves the nest to flock with its own species.

After completing their family responsibilities, the golden-cheeked warblers pack up and head south. By August, the birds have left Texas and are winging their way through Mexico to their winter refuge. Naturalists first discovered golden-cheeks in Guatemala in 1859 and found them in Texas five years later. Texans are stewards of this precious and irreplaceable biological legacy, a golden gem hidden in the scenic Hill Country. The tiny warbler is a product and symbol of the ecological richness and diversity of the rolling hills of Central Texas. Today you can see these native Texans in about ten state parks and wildlife management areas in Central Texas, several county and municipal parks, and private sanctuaries.

What will we lose if we clear the cedar brakes and drive the golden-cheeked warbler to extinction? Like a coal miner's canary testing for poisoned air, the status of the pretty warbler alerts us to a far greater danger: a destructive attitude toward the world, the source of our existence. In the complex web of life, the disappearance of one species indicates that many links in the ecological chain have been broken. If we drive the golden-cheek to extinction, we will have extinguished not just a single bird but also a vast segment of the environment. If, like children playing irresponsibly with the family heirlooms, we discard the biological treasures around us, we only demonstrate our avarice and immaturity. Like the cowbird, we will be parasites exploiting the riches of the planet, proving that we have the capability of driving nature itself into bankruptcy. Such an attitude not only destroys the quality of human life in the short run, but, if continued indefinitely, will lead to the extinction of the human species.

The Whooping Crane: Texas' Most Celebrated Visitor

Whooping crane
Grus americana

One hundred and fifty pairs of eyes strained for a better view of the 4 stately subjects in the distance. Those with binoculars and cameras leaned perilously over the rail to get a few inches closer. Suddenly all were quiet, then a simultaneous gasp. The whooping cranes had burst into the air and, with their long necks fully extended, flew directly in front of the tour boat. Their wings, stretching seven feet from black tip to black tip, beat gracefully, almost in slow motion. Clicking cameras sounded like the rumble of a distant cattle stampede. The flying foursome brought to 28 the number of whoopers we had seen on the four-hour excursion into the heart of the rare bird's feeding grounds. The captain had beached the boat twice so we could get better than a passing view. After that climactic scene, the skipper turned around and brought the satisfied tourists back through Aransas Bay to the marina.

Thousands of people visit the Texas coast each autumn to see one of the state's most celebrated attractions, the only wild flock of breeding whooping cranes in the world. In mid-October, the birds begin arriving from their summer home 2600 miles away in Wood Buffalo National Park in northern Canada. They roam the rich coastal marshes until early April, then return to their Canadian breeding grounds. In 1987, a record 134 birds, including 24 chicks, made the arduous flight to Texas.

Whoopers were never very numerous. The National Audubon Society estimates that their population probably never exceeded 1300 or 1400. The birds once wintered all along the Gulf Coast and into central Mexico. For thousands of years, they bred from Texas and Louisiana north through the central states into upper Canada. The early pioneers who encountered the five-foot white cranes named them for their trumpetlike call, which resembled the war whoop of attacking Indians. By 1870, ranchers and farmers had settled most of their range and converted prime whooping crane habitat into prime beef, corn, and wheat habitat. Whoopers last bred in Texas in 1878 near Eagle Lake, and in the United States in 1889. In 1894, a pair nested in Iowa, but no eggs hatched. By 1918, the wintering grounds of these magnificent birds had been reduced to three isolated localities. Twenty-five birds wintered in the coastal marshes in Aransas County, 16 on the King Ranch in Kleberg and Kenedy counties, and 6 in Louisiana. The White Lake marshes of Louisiana harbored an additional 12 resident birds. The last whooper left the King Ranch in 1937, and only 15 stalked the marshes around Aransas Bay by 1941. The sole bird remaining in Louisiana left the state via helicopter in 1950 to join the Aransas flock.

In 1937, Congress established the 54,829-acre Aransas National Wildlife Refuge to protect the last of this disappearing species. The marshes and shallow ponds bordering Aransas Bay and Matagorda and San Jose islands provide ideal habitat for the whooping cranes. Each family unit, usually a pair with their young, requires between 300 and 600 acres to provide enough food to satisfy their voracious appetites. Like ballet dancers in slow motion, they wade through the shallows spearing blue crabs, clams, crayfish, and other aquatic organisms. The family members feed together harmoniously, but chase intruding whoopers from their territory.

Texans get to view the whoopers from October through April, but unfortunately, they rarely see their spectacular mating dance. In the early spring just before migrating north, whoopers begin courting. They may dance occasionally during migration, but do not perform in earnest until they arrive at the 11-million-acre preserve in Canada. These graceful birds exchange bows as though they were attending a formal reception, then leap high into the air like an excited child. Wild wing-flapping, gyrations, and acrobatics accompany the stiff-legged prancing and leaping. The clarion call of the cavorting birds echoes across the marsh like a bugler sounding a charge. The dance strengthens the bond between the couples, who mate for life, and forms new pairs among maturing young.

Between May and early June, the female lays two eggs on a mound constructed of vegetation in the middle of a shallow pond. Most species of birds begin incubation after laying their full clutch of eggs so the young will all hatch the same day, but not whoopers. The female begins incubating the first egg, which hatches several days before the second. The older chick usually pecks the smaller chick to death or outcompetes it for food. Only rarely do both survive. The chicks, feasting on minnows and other aquatic and plant life, grow rapidly. By migration time in September and October, they have almost reached adult size, but they still wear their rusty juvenile plumage. The youngsters accompany their parents to the Aransas refuge and, begging for food, wade beside them through the marshes.

From 1981 to 1984, wildlife biologists followed and observed the birds along the entire 2600-mile migration route using solar-powered tracking transmitters placed on the juveniles' legs. The whoopers typically leave their nesting grounds in the southern Northwest Territories in mid-September, but may delay in Saskatchewan for several weeks to feed in wheat and barley fields. By mid-October, the family begins its flight south in earnest and may cover 450 miles a day if winds and weather cooperate. Inclement weather may cause them to lay over for a week or more. During the 30-plus days of the birds' trip, they cross the corner of Montana, the Dakotas, Nebraska, Kansas, and Oklahoma and enter North Texas. They may spend only 15 to 20 days in the air, with an average flight speed of 30 mph. While en route, the cranes feed in ponds, marshes, and grain fields, most of which are privately owned. Grains provide an important food source during migration. In pastures they turn over dried cow patties looking for insects.

The long trip between Texas and Canada poses a grave threat to a whooper's life. Hunters sometimes mistake them for snow geese, which also have white wings with black tips. The state has set a $4000 to $10,000 fine for killing a whooping crane. Power lines frequently take their toll. In 1981, one juvenile carrying a transmitter had a fatal encounter in Saskatchewan, and a yearling died near Waco, Texas, the following year. In 1983, a juvenile died in Aransas from disease, and in 1984 another crane died from a head wound, probably from a great horned owl. In 1984, a whooper at the Bosque del Apache Refuge died of lead poisoning. Since 1951, 10 cranes have died and 15 disappeared at the Aransas Refuge. At the Canada breeding grounds, the chicks must survive cold, drought, starvation, disease, and predators. Since 1977, at least 142 chicks have hatched, but only 75 lived long enough to arrive at Aransas. Oil spills, polluted water, and dredging present continual and serious threats at the Aransas refuge.

The total number of captive and wild whooping cranes reached a low of 23 in 1941. The Aransas flock of 15 contained only 2 to 3 breeding pairs. The population slowly increased at a rate of 3.6 percent per year during the next four decades. In 1972, one-fourth of the flock mysteriously died. By fall 1985, the Aransas whoopers numbered 97, and about 40 lived in captivity, most in the Patuxent Wildlife Research Center in Laurel, Maryland. In 1986, 28 whoopers nested successfully in Canada, fledging 20 chicks.

Texas can boast the only breeding flock of whooping cranes, but hopefully not for long. With only marginal success at captive breeding at the Patuxent center, biologists decided to establish a second wild flock using sandhill cranes as foster parents. Since only one of the two eggs a whooper lays usually survives, one can be removed without adversely affecting the breeding flock. The smaller sandhill cranes will incubate the whooper egg and dutifully care for the adopted young. After six years of study, biologists chose the sandhills in Gray's Lake National Wildlife Refuge, Idaho, as the foster parents. The Idaho sandhills seemed well suited for the task since they had a relatively short migration route to Bosque del Apache National Wildlife Refuge in New Mexico, and most of their stopovers occurred on protected refuges. Biologists began transferring whooper eggs to the sandhills in 1975.

By 1986, the experimental flock had grown to 37. Biologists transferred 15 more eggs from Canada that year. Unfortunately no female whoopers have survived long enough to reach sexual maturity. Some males have begun establishing territories at Gray's Lake, but the younger females have shown no interest. Instead of spending the summer together where pairs could form, the flock has scattered over parts of Idaho, Montana, and Wyoming. In 1986, biologists captured three females in Wyoming and transferred them to Gray's Lake. It is hoped that they will form pairs with the males there. Banded birds in the Aransas flock began breeding after four or five years. With the Gray's Lake population steadily growing, biologists believe that it is only a matter of time before the flock will have enough mature individuals of both sexes to begin breeding and rearing young.

The success of the programs to save the whoopers results from the collective efforts and cooperation of many agencies within the U.S. and Canadian governments and the private sector. The plight of these noble birds captures the imagination of the public. In our urbanized society, the whooping crane has become the symbol of the wildness of nature that is so rapidly disappearing and, as Aldo Leopold stated, "the symbol of our untameable past."

Reptiles

The old western movie *The Good, the Bad, and the Ugly* could as easily have been made about reptiles as about outlaws. In which category you place these scaly creatures depends primarily on your culture and your time in history. With representatives of the nearly 6000 species of reptiles occurring on every continent except the frozen Antarctic, people have lived close to reptiles since the beginning of human history. Reptiles, particularly snakes, have played a dominant role throughout the ages in mythology, folklore, and superstitions. Various snake deities reportedly guarded the treasures of the Egyptians and the secrets of the Phoenicians, and taught astronomy to the Indians. The Bible and most of its adherents curse the snake as the tempter in the Garden of Eden and consider it the symbol of evil and suffering in the world. Conversely, the American Indians, from the pre-Columbian Toltecs, Aztecs, and Mayas with their feathered bird-serpent Quetzalcoatl, to the modern Hopis with their snake dance ceremonies, have viewed snakes as symbols of the deity. The Romans associated the snake with the god of healing, and even today the snake-coiled staff remains the internationally accepted symbol of the medical profession.

Six major groups of reptiles exist in the world today: lizards, snakes, crocodilians, turtles, the tuatara, and the amphisbaenians (not to be confused with the amphibians, a separate class of vertebrates). Reptiles thrive in every inhabitable environment on the earth. These adaptable animals live from just north of the Arctic Circle to within five degrees of the Antarctic Circle. The marine iguana in the Galapagos

Islands dines on seaweed beneath the ocean surface, while a lizard in Nepal scampers across rocks on 18,000-foot mountaintops. Reptiles thrive in inhospitable deserts, humid swamps and marshes, boreal, temperate, and tropical forests, scrublands, and prairies. They burrow through the soil, crawl legless across the ground, climb trees, cling to vertical surfaces, plod slowly along, race swiftly away, and swim in both salt and fresh waters.

Unlike birds with feathers and mammals with fur, reptiles have no single distinguishing characteristic. Scales, which provide a tough, waterproof shield, cover their bodies, but fish, birds, and some mammals also have scales. All reptiles are ectothermal, or cold-blooded, a misnomer since their blood is not cold. They cannot maintain a constant internal temperature like birds and mammals, but must depend on the environment to warm their bodies sufficiently for the chemistry of muscle action and digestion to function. Reptiles must seek shade to prevent overheating, and most prefer an 80-to-100-degree F range.

With approximately 3000 species, the lizards make up slightly more than half of the living species of reptiles. Though they typically have well-developed legs, a long tail, sharp teeth, ears, and keen eyes with eyelids, many species depart radically from this generic description. The bodily features of each species enable it to survive in a particular environment. The size and shape of its body, legs, and toes, the acuity of its sight, hearing, and other senses, and its diet, reproductive strategy, and behavioral habits match it to a specific niche and habitat. The toes of a gecko have pads with as many as one million microscopic bristles with suction cups. The specialized toes enable the gecko to climb vertical surfaces and even hang upside down to catch the insects it eats. Many lizards living in sand dunes have long, fringed toes to help them scurry across the loose sand. A lizard's long legs give it speed and also allow it to elevate its body off the hot desert sand and rocks. A following section describes the extreme modifications of the Texas horned lizard that uniquely adapt it to feed on its primary food, ants.

Every child, or adult, who has tried to catch a lizard has learned that most have tails that pop off if grabbed. The broken appendage wiggles violently, startling or confusing the attacker while the lizard scampers to safety. A special set of vertebrae, with a muscle to clamp off the blood flow, allows a lizard to sacrifice its tail without harm. Amazingly, the tail regenerates, though the new growth has no vertebrae and is never as long. Some lizards have other defenses equally as bizarre as a detachable tail. The horned lizard, besides having a spine-covered body, squirts blood from its eyes, a chuckwalla jams itself in a crevice by swelling up, and a chameleon changes colors. Most lizards depend on camouflage and speed for their defenses.

Perhaps no group of animals has aroused the imagination and ire of humans more than the serpents. Only in Ireland, New Zealand, and Greenland do people live free from the fear of snakebite. Unfortunately for the rest of the clan, numbering about 2700 species, the poisonous snakes have created such a bad public image that most people indiscriminately fear and kill all snakes, at least in the Western world. Many Asian cultures consider them a gourmet food item and an important pharmaceutical ingredient for medicines.

Snakes have adapted to a greater variety of environments than any other cold-blooded animal. Besides the terrestrial habitat, some live in the oceans and fresh waters, in trees, or underground. Beyond the remarkable adaptations in body shape and scale structure that enable some snakes to swim and others to climb or burrow, a snake's body represents an extreme example of specialization. When survival pressures began molding the lizard predecessors of snakes into slender, tube-shaped creatures that could hide in cracks and tunnel underground, a number of internal and external modifications were required. During 60 million years of underground existence, snakes lost their legs, a hindrance to burrowing creatures, as well as their ears and keen vision, unnecessary senses in a subterranean world. Developing a long, flexible body required extensive modifications of the spine, ribs, and skull. Some snakes have as many as 400 vertebrae with movable ribs. The tubular body also required a rearrangement of internal organs. One lung disappeared and the other elongated. The respiratory system, including a modified trachea, occupies three-quarters of the snake's length. Today only boa constrictors and sunbeam snakes have a functional left lung. Snakes have an elongated stomach and no bladder, with the other organs staggered through the body. These slender creatures store the food reserves necessary for reproduction and hibernation in special fat bodies in their tails.

Snakes have developed various methods to catch and subdue live prey, the diet of all serpents. A constrictor wraps itself around its victim and prevents it from breathing. These snakes suffocate, not crush, their prey. Venomous snakes have a special poison-delivery system to inject fast-acting toxins into animals. The poisons combine a complex mixture of modified digestive enzymes specialized to attack the blood, internal organs, and nervous system. Other snakes simply chase down and overpower their prey.

No reptile can chew its food into small pieces. For a snake to swallow a creature that may be two or three times its diameter requires extreme anatomical modifications. A highly specialized skull allows a snake to flex and rotate the two halves of its lower jawbone independently and pull its food inward with its backward-facing teeth. The lower jaw completely unhinges from the upper, and the neck and body skin stretches to allow a snake to swallow food. A reinforced windpipe prevents the snake from suffocating while engulfing large prey: a 25-foot captive python once swallowed a 74-pound goat in two hours. The astounding adaptations of snakes make them among the most biologically interesting creatures alive.

The third largest group of reptiles, the turtles, swim the open seas, rivers, and lakes, and trudge through the deserts, plains, and forests of the world. Turtles, with their unique body covering, have a radically different appearance from all other reptiles. Their shells provide protection, and in some species are hinged to close tightly. Contrary to folktales, a turtle cannot crawl out of its shell, which is fused to its backbone. The 250 species living today vary from fist-size box turtles to the giant land tortoises of the Galapagos Islands and eight-foot leatherback sea turtles. Scientists consider the land tortoises the most intelligent reptiles. Turtles live longer than any other vertebrate, with an American box turtle holding the record of 123 years.

The crocodilians, including alligators, crocodiles, caymans, and gavials, comprise another order of reptiles. These semi-aquatic creatures, with 29 species worldwide, thrive in the tropics and subtropics. The American alligator, described in a following section, has only one other member in its family, the Chinese alligator, while the gavial of India is in a family of its own. The other species belong to the crocodile family.

Crocodilians of all sizes, from the 4-foot dwarf species to the giants, come well equipped for both killing live food and defense. With sharp teeth and powerful jaws, designed for crushing and ripping bite-size chunks from dead prey, even small specimens can easily inflict serious injury. In general, crocodiles have a temperament that makes the American alligator gentle in comparison. The Nile crocodile of Africa and the saltwater crocodile of Asia reach 20 to 25 feet in length, weigh 2000 pounds, and have a disposition intolerant of people. These monsters habitually prey on humans and are greatly feared by people living in areas where they are abundant.

Two other groups of reptiles have few members and are rarely encountered. The amphisbaenians are snakelike creatures with no external ear openings; only a few members have visible limbs. The worm lizard, the only member in North America, lives in central Florida. This unusual subterranean animal has no eyes or legs and closely resembles an earthworm, one of its primary food items. Tiny scales cover its 12-inch body, and it has a lizardlike head. The last group of reptiles has one surviving member, the tuatara of New Zealand. This 2-foot lizardlike relic of antiquity has a bony cranial structure dating back to the beak-head reptiles of 150 million years ago. The tuatara, as well as some lizards, including the fence lizard of Texas, has a vestigial third eye on the forehead, complete with lens, cornea, retina, and nerve connections to the brain. The organ probably is sensitive to day length, which influences reproductive functions, and helps the animals regulate their exposure to the sun.

Reptiles were the first terrestrial vertebrates to develop after some ancient species of fishes left their watery home and evolved into amphibians. The key to the success of reptiles, the factor that propelled them beyond the dependence on water of their amphibian predecessors, was the development of an egg covered with a leathery, but porous, shell. No longer did reptiles have to return to the water to lay their eggs—they could bury them in the soil and let the earth and sun incubate them. Also, the embryo could develop fully in the protected egg and skip the larval, or tadpole, stage. Today some lizards and snakes keep the eggs inside their bodies until they hatch, giving birth to live offspring.

Fishes first adapted to life on dry land during the Devonian Period. Fossil remains of 400-million-year-old fish show unmistakable signs of lungs and thick, fleshy fins that would have allowed them to crawl from one pond to another. Today's walking catfish and lungfish use the same methods to survive droughts. Scientists consider the fossilized fishlike creatures to be links between fish and the first amphibians. The amphibians adapted to life on dry land, and some eventually evolved into reptilelike animals. By Permian time, 250 million years ago, fossils record a great abundance and variety of reptile forms. Reptiles dominated the landscape for the next 190 million years, to the close of the Cretaceous period: dinosaurs, varying from the size of cats to 90 feet in length, roamed the earth; pterodactyls, some with 40-foot wingspans, soared through the sky; while 50-

foot aquatic reptiles swam the depths of the oceans. Fossils record a line of reptiles that exhibit many mammalian characteristics and are considered predecessors to all living mammals. Another series of fossils reveals reptilelike creatures with feathers, representing the ancestors of today's birds, which still retain many reptilian features, including scales on their legs.

Paleontologists trace today's reptiles back through the fossil records to their ancestors in the past. Turtles are the old-timers of all reptiles. They saw the dinosaurs come and go, and today remain basically the same as they were 275 million years ago. The oldest lizard fossils date back to the Triassic period 190 million years ago. Snakes first appear in the fossil record about 135 million years ago.

Predation pressures from small, agile dinosaurs probably forced many lizards to live in rock cracks and crevices and to burrow underground, habitats that favored slender, flexible bodies. In a subterranean habitat, legs, eyelids, and ears were not as important as a tube-shaped body with the necessary internal organ modifications. The extreme restructuring of the lizard's body into the form of a snake indicates intense pressure either to change radically or to perish. Snakes adopted a completely subterranean lifestyle until about 65 million years ago, when a great extinction eradicated all large terrestrial, marine, and flying reptiles. The extinction theory currently supported with the most evidence proposes that a large meteor struck the earth and that the massive amount of particulate matter displaced into the atmosphere obscured the sun for a number of years. Without sunlight, plants and the animals that depended on them died. With the major predators gone, the snakes emerged and proliferated. At least one group of modern snakes retains evidence of their ancestral form. Boas and pythons have clawlike vestigial hind limbs on either side of their vent.

Texas, with its great environmental diversity, is well populated with reptiles. The state hosts about 110 species and subspecies of snakes (far more than any other state), sixty-one species and subspecies of lizards, and thirty-five species of turtles, as well as a healthy population of alligators. Unfortunately the taming of Texas, and the rest of the world, is having a tragic effect on reptiles. Habitat lost to farms, ranches, and cities and the use of insecticides and other poisons have placed 40 species of reptiles in the threatened and endangered classification. The official list of threatened and endangered wildlife, compiled by the Texas Parks and Wildlife Department, includes about 36 percent of the snake species, 11 percent of the lizards, and 28 percent of the turtles.

In the next seven sections, you will discover more about the poisonous snakes in the Lone Star State, where they live, and the latest medically approved methods for treating snakebite. You will learn the amazing adaptations of the strangest lizard in Texas, read about a 25-year effort to save a rare turtle, and find the success story of one reptile that was saved from the brink of extinction.

The Alligator: Gator-Aid in Texas

American alligator
Alligator mississippiensis

Children pressed their noses against the restaurant windows and adults left their lunches to watch the spectacle. I felt like a Roman gladiator facing a hungry lion while the masses cheered, favoring the lion more than me. Right on cue, Charlie the alligator swam to the pond's edge, and waited with his head raised halfway out of the water and his mouth gaping menacingly. I unwrapped the chicken. The kids screamed "Jaws!" and the adults gasped when I tossed it a foot to the right of the gator's mouth. With a reflex born from millions of years of instinct, the ten-foot gator struck the food with lightning speed. With two crunches, Charlie's powerful jaws positioned the meal. Then he swallowed it whole and settled back into the water. The show wasn't long, but it satisfied the customers and sold hamburgers.

Alligators trace their ancestry back to the distant age of reptiles when flesh-eating creatures roared and roamed through the swamps. They have remained unchanged for 35 million years, sufficient time to tune their survival instincts to machinelike precision. A Port Arthur hunter who almost lost his retriever to an alligator called them "the meanest killing machines that crawl the face of the earth." To most people, they seem as threatening as the movie monsters that terrorize cities.

In times past, alligators reached 19 feet in length and slid into southern swamps by the millions. In the nineteenth century, hunters killed 25 million in Florida alone. Many considered these giant reptiles big game animals and searched the marshes for the largest trophies. Midway into the twentieth century, alligators had disappeared from most of their range and, like the passenger pigeon, seemed doomed to extinction. Only about half a million had survived the 150 years of slaughter and habitat destruction. Today the American alligator ranges throughout swamps and marshes from the mouth of the Rio Grande to North Carolina. In Texas, most live in the coastal marshes, but a few survive along the river courses into Central Texas. About 15 find refuge in the Highland Lakes from Town Lake in Austin to Lake Buchanan.

These vanishing relicts from prehistory received federal protection in 1967 under the Endangered Species Act. By 1980, the Texas population had grown to 68,000, numbering about 140,000 by 1985. Protected from their primary predator, humans, alligators increased rapidly until Texas, like Florida and Louisiana, classified them as threatened and allowed controlled harvesting. In 1985, Texas hunters killed 747 in a closely supervised hunt.

The rapid increase from near extinction to abundance shows us how well alligators are adapted for survival. These silent hunters patrol the waterways consuming fish, mammals, birds, and carrion; they catch floating animals by swimming underneath them, surfacing, and lashing sideways with jaws open. The mighty jaws, studded with teeth, can crunch a turtle like a boiled egg, snap bones like match sticks, and tear an animal into shreds. When the food is too large to swallow whole (alligators can't chew) they rip off a large chunk by biting down and violently jerking their head back and forth, like a shark.

Although alligators eat smaller creatures, many of these same animals depend on the large predator for their own survival. Marshes periodically dry up, and the many fish, amphibians, turtles, birds, and mammals dependent on water would perish if numerous small ponds did not exist. The lifesaving ponds are not just natural depressions, but "gator holes." Alligators use their mighty tails like bulldozers and thrash out a wildlife oasis 10 feet deep and 40 feet in diameter. Each gator has its private pool and a den that may reach

10 to 15 feet into a mud bank. Despite their reputation, alligators play an imperative role in the ecology of the marshes, and all wildlife suffers when they disappear.

Alligators survive by eating anything they can catch and, when food is scarce, by going long periods without food. Like all reptiles, an alligator's internal temperature depends on the surroundings. These ectothermic, or cold-blooded, animals require a water temperature of about 80 degrees F to warm them up enough for their digestive system to function. For years, the ten-foot gator at the hamburger restaurant ate 1, sometimes 2, chickens a week from late June through September, and nothing during the rest of the year. His yearly diet consisted of approximately 15 fryers, an amount of food easily obtained in his natural marshy habitat. In the winter, alligators retreat to their dens and hibernate. I could drag Charlie out of the frigid water by his tail to wash his pool, a feat that would have been suicide in the summer. Since the water in his pool never froze, Charlie easily survived the Austin winters, even when the air temperature dropped well below freezing.

After reaching several feet in length, alligators have little to fear from other animals. The reptile's large size, sharp teeth, and powerful jaws persuade neighbors to avoid any confrontation. With no enemies and an easily satisfied diet, alligators have thrived in the southern United States for countless millennia. Another adaptation for survival, the ability to produce many offspring, has enabled their population to rebound after reaching the dangerously low numbers of recent times. In the spring, the throaty bellow of the male echoes through the marshes like a car with a rusty muffler. A bull alligator fiercely repels intruding males from the female, and violent battles may churn a tranquil bayou into a bloody froth.

The female constructs a nest mound of mud, sticks, and vegetation three feet high and ten feet wide. She lays from 30 to 60 hen-size eggs and covers them with more vegetation, which keeps the eggs warm as it decomposes. Unlike most reptiles that abandon their eggs, the mother alligator dutifully guards the nest during the nine or so weeks of incubation. When the eggs begin hatching in late August and early September, the babies make noises that tell the mother to dig them up. Sometimes the mother even takes the babies to water in her mouth. She protects her young for several months, but most fall prey to raccoons, herons, gar, and other alligators, which find the eight-inch babies tasty morsels.

Young alligators eat tadpoles, minnows, crayfish, insects, and other small animals. I once had an 18-inch gator that I hand-fed minnows, which it grabbed with the typical sideways lashing attack with its head. After a few nasty encounters with its sharp teeth, I decided it could catch its own fish in the small pool that was its temporary home. Alligators grow about a foot a year for their first six to ten years, then about an inch a year. Most gators in the wild today measure between 6 and 10 feet and weigh 100 to 200 pounds. A 12-foot gator could weigh as much as 450 pounds. Even with today's limited hunting, an alligator will probably never reach the 20-foot size of the creatures that glided through the swamps of the past.

People boating or swimming generally react with fear at the sight of a nearby alligator in the water. Though retiring by nature, any alligator represents a potential threat, whether on the land, in shallows, or in deep water. An alligator attacks by thrashing sideways, and any person or animal within striking range, about half the length of its body, cannot move fast enough to escape the jaws or tail. The powerful animal uses its tail like a bludgeon, and it can whip it sideways with enough speed and force to break a leg. Alligators appear docile and sluggish, but if the occasion arises, they can run as fast as a human for short distances. Wild alligators, however, seldom will attack a human, unless cornered. Their reptilian brain may lack intelligence, but they are smart enough to recognize a threat. Gators usually avoid any animal as large as a human. I have seen fishermen standing in shallow lakes in Florida with alligators only a few yards away, each ignoring the other. Injuries sometimes do occur when people feed alligators regularly, and the animals lose their fear of humans. An alligator accustomed to eating food thrown from the water's edge is expecting a meal when a person or pet jumps into the water.

In 1979, a University of Texas student illegally collected alligator eggs from an area on the Texas coast destined for development, hatched them in his apartment, and released 316 in the lakes around Austin. Despite the media hysteria (with headlines such as "Guess Who's Coming to Dinner—You!") alligators did not eventually infest the lakes. Raccoons, birds, and gar probably ate most, if not all, of the hatchlings. An 18-inch gator did appear on an Austin newspaper columnist's desk, which caused quite a commotion in the newsroom, but it was too large to have come from the clandestine eggs. A few adult alligators do live in the Highland Lakes, but they rarely are seen.

Alligators and their ancestors have a 200-million-year legacy in Texas. In the distant past, when West Texas was a swampy marsh, giant alligators basked in the sun and slithered into the tepid waters. Archaeologists in Big Bend National Park unearthed the 6-foot fossil skull of a gator that would dwarf today's critters: the prehistoric creature measured at least 45 feet long. Alligators have survived massive loss of habitat by planetwide climatic changes, advancing and receding oceans, and ice ages. Though human hunters and developers nearly drove these adaptable reptiles to extinction, the future of the alligator now seems secure, thanks to the efforts of conservation groups and the protection of state and federal agencies.

The Copperhead: The Reclusive Viper

Southern copperhead
Agkistrodon contortrix contortrix
Broad-banded copperhead
A. C. laticinctus
Trans-Pecos copperhead
A. C. pictigaster

If snakes could talk, I know one copperhead that would have yelled, "Hey, watch your step, lady!" Our group had hiked deep into the Big Thicket for several hours when one lady stepped right on the middle of the perfectly camouflaged snake. She screamed and jumped straight into the air, and the snake slowly squirmed away, unharmed. When we all gathered around for the most exciting sight of the day, the aroused serpent coiled in its defensive position and vibrated its tail. In the loose leaves, the buzz sounded an unmistakable warning, like that of a rattler. A few feet away by a rotting log lay another copperhead, silently observing the predicament of its companion. Thanks to the unaggressive temperament of these retiring serpents, no one was bitten that day.

A high school buddy of mine was less fortunate. While in the woods, he saw a colorful piece of rope on the ground and reached down to pick it up. Suddenly the rope came alive and bit him on the thumb. The mottled white and brown pattern of a copperhead conceals it so well that it becomes almost invisible on the leaf-littered forest floor. The bite my friend received was frightening and painful, though not serious, but he almost lost his hand from the tourniquet his terrified friends wrapped around his wrist.

Despite their venomous threat, copperheads are one of the most beautiful snakes in North America. Bands of rich golden tan, the color of fallen oak leaves or pine needles, alternate with beige to off-white bands. The combination of rich earth colors and lighter markings camouflages the snake perfectly in the dappled shadows of its woodland environment. Its variegated pattern and the habit of lying motionless instead of slithering away when threatened are a copperhead's best defense, and offense. The cryptic coloration enables the snake to remain undetected when danger approaches, or while waiting for prey to come within striking range.

In East Texas, where they are the most common venomous snakes, copperheads inflict the majority of poisonous bites, yet fatalities seldom occur regardless of treatment. A copperhead's venom has only 66 percent of the lethal potency of a western diamondback rattlesnake's, and 33 percent of the blood-destroying capacity. With fangs less than 5/16 inch long, copperheads usually inject their poison into the skin layer, not the deeper tissues and muscles. A bite from a large copperhead can deliver only about 10 to 15 percent of the poison necessary to kill an adult human; however, the snake sometimes strikes repeatedly. A bitten hand or arm may swell to twice its size, and severe poisonings may cause loss of digits or limb. Most permanent injuries from copperhead bites result from ill-advised tourniquets and the cut-and-suck treatment instead of from the effects of the venom.

The three subspecies of copperheads in Texas favor riparian woodlands with a dense leaf litter. The southern copperhead of the eastern third of Texas and the southern states lives primarily in pine and deciduous forests. It is especially abundant in wet, low-lying areas, and may reach densities of six to seven per acre in rich environments. The broad-banded population extends approximately from Dallas south to Victoria and west to Del Rio and Abilene, and northward to Kansas. This snake favors post oak woodlands with a sandy soil covered with leaves, and the oak and juniper savannahs and rocky ledges of the Edwards Plateau.

The Trans-Pecos copperhead, the rarest of the three, lives near permanent water in the desert, usually in association with cottonwood or tamarisk trees, though it has been seen in the shrub desert miles away from water.

The optimum body temperature for a copperhead is between 70 and 80 degrees F, which it regulates by basking in the sun when the air temperature is cool, and avoiding the heat of the day in the summer. A copperhead caught out during a torrid West Texas day could easily exceed the fatal body temperature of 104 degrees. In the spring and fall, they hunt at dawn and dusk, emerging later in the evening as the summer temperature increases.

With excellent night vision for a snake, heat-sensing facial pits, and a sophisticated venom, copperheads are superbly adapted for hunting small animals in the dark. These opportunistic feeders mainly catch white-footed, harvest, and deer mice, but in the late summer they often gorge on cicada nymphs. Insects comprise about 20 percent of their annual diet. Lizards, small snakes, frogs, toads, and skinks also find their way onto the copperhead's varied menu.

Unlike snakes that stalk or chase their prey, copperheads ambush their victims. These sluggish 20-to-30-inch snakes expect curb service and lie in wait on an animal trail until their food comes to them. Their expenditure of energy for an evening's meal might be as little as a 5-inch strike. With their acute smell, they sometimes find the nests of mice and ground nesting birds.

The juveniles use another technique to find food. Copperheads naturally lie perfectly still so their camouflaged coloration hides them. The young hold the bright yellow tip of their tail straight up and slowly wave it back and forth. Frogs and other small critters mistake the tail for a meal, and become the meal themselves. Whether the snake intentionally waves its tail as a lure or it is a nervous response to waiting for the strike is undetermined, but the ploy works. Young water moccasins use their tail as a lure in the same manner.

With stout bodies adapted to a sedentary life, copperheads seldom wander far. In a rich environment, females may spend their entire life within 1 or 2 acres, while males range through 8 to 25 acres. A group of marked young in a North Texas study plot strayed less than 100 yards during the first year. Occasionally these terrestrial snakes crawl into grapevine-covered trees to avoid high water and to search for food. During the spring and fall mating seasons, the males often abandon their nocturnal habits and search for females during the day.

Most females reproduce every other year because of the time required to store the necessary fat reserves. From 5 to 14 young are born alive from late July to September and early October. The seven-to-ten-inch babies begin life with a full load of venom and aggressively strike at any disturbance, which undoubtedly saves many from hungry predators. Moles reportedly burrow underneath the young and snatch them from below, while skunks, opossums, and larger snakes search through the leaf litter for the morsel-size tidbits. Adult copperheads have fewer foes, though hogs, with a thick skin and layer of fat, devour them with impunity. Birds of prey probably seldom pounce on the well-hidden snakes, yet a great horned owl did attack one in a herpetologist's outdoor pen. In an unusual outcome, the snake killed the owl.

Copperheads often live in wooded suburban areas, especially in East Texas. They survive in overgrown vacant lots, in brush piles, and along creeks until they are discovered and killed, or their habitat is cleared. Southern copperheads were the most common land snake I saw while growing up in a wooded area north of Beaumont, and I can recall many encounters with the colorfully banded snakes. A particularly favorable microhabitat, such as a good winter den, a well-suited basking area, or a location with abundant food, may attract a number of snakes, including copperheads. Two stacks of corrugated iron siding at an abandoned farm in North Texas sheltered a dozen broad-banded copperheads.

Most snakebites occur not in the wilds of nature but in backyards of homes in the most populous cities in the state. Children and teenagers, the largest group of victims, receive 46 percent of the bites. You can go a long way toward preventing snakes from living in your yard by removing piles of lumber, brush, rocks, and rubbish that might harbor both the serpents and the rodents they eat. Clearing dense vegetation eliminates cover and hiding places for snakes. These precautionary steps provide much better protection than the chemical snake repellents on the market. Most are either completely ineffective or too toxic to be used around children and pets, or where runoff would pollute streams and lakes.

The Coral Snake: "Red Touch Yellow, Kill a Fellow"

Coral snake
Micrurus fulvius tenere

Most of the phone questions I received during my tenure with the Austin Nature Center concerned snakes, and "What kind of snake is this?" was the most common. One person wanted to know how to identify a coral snake. I quoted the rhyme learned in my childhood, "Red touch yellow, kill a fellow," referring to the arrangement of the snake's stripes. Before I could finish with "Red touch black, venom lack," the caller sympathetically replied, "Oh, I'm so sorry the fellow died!"

I remember the first coral snake I saw, in the woods along Armand's Bayou near Houston. Fascinated, I followed the snake crawling nonchalantly through the thick leaf litter, and repeated "Red touch yellow, kill a fellow" like a child trying to memorize a lesson. Finally, just as the beautiful but deadly creature disappeared under a rotten log, I was convinced my eyes weren't deceiving me.

Though common throughout the eastern two-thirds of the state, from Dallas to Brownsville and west to the Pecos River, coral snakes seldom come to our attention. These gaily patterned serpents prefer to hide their vibrant colors in woodlands with thick leaf litter and in areas with abundant outcroppings of rocks. They live in the rocky streambeds of Central Texas, chaparral thickets of South Texas, the Cross Timbers woodlands, and East Texas Piney Woods. Thick terrestrial cover provides both a hiding place for the snake and the habitat for their reptilian prey.

Given the proper cover, coral snakes coexist contentedly in suburban settings, particularly in parks and wooded areas. They venture forth in the morning and evening searching for their favorite food: other snakes. Once, on the sidewalk of an Austin park, I found one that had swallowed all but the last four inches of a rat snake. Other reptiles in their diet include rough green, earth, and flat head snakes, as well as other corals and an occasional skink.

A coral snake kills its prey with the most potent venom of any North American serpent. It is about eight times more toxic than western diamondback rattlesnake venom, and it equals that of most cobras. Between 5 and 10 milligrams will kill an adult. The snake's poison attacks the central nervous system, unlike pit viper venom, which is composed primarily of digestive enzymes that destroy tissue. The prey of a coral snake dies from respiratory and cardiac failure. With small creatures, death is rapid so the snake can consume its victim, but with humans fatal symptoms may take hours to develop. One teenager appeared unaffected by a bite, but died six hours later.

A San Antonio zookeeper bitten just momentarily on the back of his hand experienced immediate intense pain, which lasted for 24 hours. He was treated with antivenin and remained in intensive care for three days but recovered fully. He suffered none of the muscle damage and extended period of weakness usually associated with pit viper bites. Because of the sudden and sometimes delayed failure of the central nervous system, doctors recommend administering the venom antidote immediately, whether or not symptoms occur. If bitten, wrap an elastic band between the bite and heart, splint the limb, and seek immediate medical attention. Fortunately less than 40 percent of coral snake bites result in serious poisoning, partly because of the snake's tiny fangs.

A coral snake seldom strikes unless provoked, and practically all bites occur to people handling a snake. Though some corals allow repeated handling without biting, most will strike immediately if treated roughly. These serpents do not have the long, hollow hypodermic fangs of rattlesnakes and other pit vipers: the poison flows down narrow grooves on the front surface of their ⅛-inch fangs. A membrane covers the peglike fang providing a closed passageway for the flow of venom.

The choice of prey, behavior of the snake, and potency of its venom are all interrelated. A coral snake's lethal venom enables it to dine on snakes that it otherwise could not subdue and kill. Since snakes cannot chew, they must attack prey small enough to swallow whole; and since they eat freshly killed animals, they must rapidly dispatch their victims. Unlike a rattler, which strikes, injects its venom, and withdraws instantly, coral snakes usually bite and hold their prey tenaciously—thus, the scientific name *tenere*. The biting grip enables as much poison as possible to trickle into the victim. A test bite on the finger of a welding glove showed numerous scratches from the twisting, biting action of the snake.

Scientists have long speculated about the advantages of a coral snake's brilliant colors. One popular theory proposes that the bright colors and distinctive bands warn other animals of the snake's venomous defense. Similar warning coloration has been demonstrated in other animals, such as wasps. To learn to recognize the colors as a warning requires a predator to attack the snake, get bitten, and survive the ill results. A predator can learn to avoid animals having mild venom, but not those whose bite kills outright. If the attackers never survive, no learning takes place and the bright colors cannot serve as a warning.

Another theory suggests that the bright colors actually camouflage the snake. In moonlight red appears gray, and the bands look like mottled shadows. Coral snakes, nonetheless, forage primarily during the day, not night, so they would be glaringly apparent to animals that could discern colors. However, most of the mammals that kill and eat snakes—armadillos, opossums, moles, and raccoons are color blind, so the colorful pattern may in fact help them blend in with their environment.

The harmless milk and scarlet snakes, which also have a combination of red, yellow, or black bands, traditionally have been considered to mimic the deadly coral snake. Supposedly animals would mistake them for the coral snake and leave them alone. But tests show that snake-eating animals rarely have any inhibitions about attacking corals, supporting the idea that the colors do not act as warning beacons. More likely, these species independently developed the patterns and colors because they helped camouflage them from color-blind predators.

An alternate theory speculates that coral snakes themselves mimic less toxic snakes, ones predators could learn to avoid without being killed. Most of the more than 30 species of coral snakes occur in the tropics, and some are less virulent than our species. In the distant past, ancestors of the North American species might have lived farther south, where they could mimic less poisonous relatives, and then gradually expanded their range to the present distribution.

Regardless of the evolutionary reason, coral snakes are one of the most colorful and easily recognized snakes in Texas. Narrow canary yellow bands separate the broad jet black and red bands that encircle their bodies. The red bands usually have mottled black spots giving a dusty appearance. The similarly colored milk snake has red, yellow, and black bands, but only the coral snake has yellow touching the red—hence the often repeated rhyme. Additionally, only the coral snake has bands that completely encircle its body.

A coral snake's head and tail are colored identically. Both have a black tip followed by a series of yellow and black bands. When in danger, the coral snake frequently uses its tail as a threat to frighten predators. It covers its head with its foreparts, sticks up its tail in a menacing fashion, and snaps it forward simulating a strike. This posturing both protects the snake's head from attack and apparently deters the aggressor. Tests show that many animals retreat when they see the tail waving.

While most people think of coral snakes as small serpents, the adult females average about 26 inches and the males 24 inches. The record, 1/4 inch shy of 4 feet, was captured in Brazoria County in 1958. Most of the large corals are females. They lay between three and five leathery white eggs from May through July. The 6-to-7-inch, pale-colored babies hatch about the size of earthworms. Within one week, the young shed and sport the brilliant colors they will wear for the rest of their lives.

The Cottonmouth: Beware the Gaper

Western cottonmouth, water moccasin
Agkistrodon piscivorus leucostoma

Just as West Texas has its legends of giant rattlesnakes, East Texas abounds with stories about the cottonmouth, or water moccasin. The sight of this serpent strikes fear in the hearts of outdoor adventurers throughout the eastern half of the state. Most hunters in the Piney Woods and coastal marshes and recreationists on the region's many lakes have personal tales of their experiences with this fearsome reptile.

The large-bodied, dark-colored water moccasin only partially deserves its reputation as a belligerent snake. Most would rather avoid people than fight, but like any cornered animal, it will defend itself to the death. A cottonmouth threatens by holding back its head and opening its mouth wide with fangs erect. The gaping posture and the white interior of its mouth give it the common names "cottonmouth" and "gaper" and the scientific name meaning "forward-fanged, fish-eating, white-mouthed serpent."

Water moccasins usually hunt at night and eat frogs, minnows, tadpoles, and other aquatic creatures, as well as a variety of terrestrial animals. They kill small mammals with their venom and often have to track down and find the dying animal. The aquatic habitat and hunting behavior of the water moccasin have engendered a number of popular folktales. Since moccasins are accustomed to feeding on freshly killed animals, a stringer of fish looks like an easy dinner. Night fishermen always advise, "Never shine your light on a water moccasin or it will crawl right into your boat!" So what do you do if you see one? Do you turn off your light and wonder where the poisonous creature is? I never met a fisherman content with letting a moccasin swim around his boat in the dark.

Every boater in the state has heard the folktale about the

agonizing death of the water skier who fell into a nest of water moccasins. The snakes attacked him with a fury, and with more bites than could be counted, he died before reaching the hospital. The story changes little through the decades, yet such an incident has never been documented by newspapers or medical records. According to another common myth, water skiers are perfectly safe because the snake cannot open its mouth underwater to bite or it will drown. Tell that to the fish that water moccasins voraciously consume.

The story of the attacked water skier may have been born when someone saw several snakes along a lakeshore. Locations with abundant food, such as shallows where small fish cannot escape, attract water snakes, and all of the 12 species of nonpoisonous aquatic snakes in Texas are considered cottonmouths by most people. Water-filled depressions in a creek bed or rocky lakeshore provide a natural refuge for aquatic critters, including snakes. I once saw five snakes, none poisonous, in a three-foot pothole when the Pedernales River was running low. Snakes may congregate at a favorable spot, but they do not "nest" in mass like some birds and herding animals.

Despite their name, water moccasins are only semiaquatic, and may forage a half-mile from water, hunting in fields and woodlands for rabbits and rodents. Like other pit vipers, they have heat-sensing organs located in a tiny pit on either side of their nose to help them accurately strike warm-blooded prey. A high-speed film of a moccasin striking a mouse revealed that the fangs were in the animal's body only $^1/_{50}$ second, and the snake could adjust the angle of each fang independently while striking. When hunting cold-blooded fish and amphibians in murky water, cottonmouths strike blindly at any ripple or movement. Research showed that fish and amphibians comprise one-third to one-half of the snake's diet, with reptiles making up nearly one-third and birds and mammals one-third. Cottonmouths readily devour other snakes, even poisonous ones. This preference may explain why other water snakes are rare in areas with an abundant population of moccasins.

Most moccasins display little fear of humans, but although they gape and display fearlessly, few people are bitten. Less than 7 percent of the hospitalizations from snakebite in Texas result from the water moccasin. I can attest to their reluctance to bite. As a junior high student, I stepped on one in the schoolyard without the snake striking. I once sat within a foot of one, water skied over one, and, following eight other hikers along a trail, stepped over one coiled among tree roots. Cottonmouths not only show less of a propensity to strike than rattlesnakes, but their venom is about 30 percent less toxic than the western diamondback's. Though they can inflict debilitating and permanent muscle damage, less than one person a year in the United States dies from being bitten.

An adult cottonmouth has a grayish black back with often indistinct crossbands, which makes them difficult to distinguish from the nonpoisonous blotched, yellowbelly, and diamondback water snakes. The large, triangular, flat-crowned head with vertical cat-eyed pupils provides the best field mark for the cottonmouth; however, you may prefer leaving a close inspection of a live snake's eyes to a snake charmer. No other aquatic snake vibrates its tail and gapes motionlessly when threatened on land. In the water, a cottonmouth swims with its body on the surface, while other water snakes generally swim with only their heads above water. Juvenile water moccasins have a distinct pattern of crossbands and a yellow-tipped tail. Their tail acts as a lure to trick minnows and frogs. A minnow sees the flicking tail and, thinking it is a juicy worm, rushes not to its dinner but the snake's.

Between August and October, the female cottonmouth gives birth to 3 to 12 live babies. The 7-to-11-inch young must fend for themselves immediately, and avoid being eaten by other snakes, birds, and largemouth bass. By November, the moccasins begin denning up for the winter, choosing rocky slopes in the western part of their range and crayfish and rodent holes along the coast. They hibernate until March or April, although they often emerge on warm days. Cottonmouths and their close relatives, the copperheads, are some of the first snakes to become active in the spring.

In Texas, the western cottonmouth ranges west to San Angelo, south to Corpus Christi, and north to Wichita Falls. It has a patchy distribution on the western edge of its range, being rare on the Highland Lakes of the Colorado River, and absent along the Brazos River around Waco. On the Edwards Plateau and in North Texas, they live in woodlands along river bottoms and lakes. Water moccasins do not occur in South Texas, the Panhandle, and the Trans-Pecos. Their range extends east along the Gulf Coast to Alabama and north to southern Missouri.

When food is abundant, cottonmouths thrive in large numbers, especially along sloughs, bayous, and irrigation canals in the Coastal Plains. One observer near Corpus Christi recorded 30 moccasins in 250 yards. As a boy growing up in Beaumont, I considered hunting moccasins along rice canals a great sport. The thick-bodied snakes would stretch out on willow limbs a few feet above the water. The challenge was to sneak up close enough for a shot before the snake dropped into the water. Sometimes the snakes basked in the sun on the levee and dashed for the water straight toward me, which only added to the excitement.

I now find pleasure in admiring the creatures I once shot, both for their important ecological role and for their value as fellow inhabitants of this planet who have a right to live their own lives. Many people disagree and place a value on animals, and all of nature, relative only to their needs and likes. Whether this self-centered view of life is an artifact of our consumer-oriented society or a survival instinct is irrelevant. Though in certain situations hunting benefits an animal population or is necessary for human safety, the same viewpoint that justifies the needless killing of animals for personal pleasure also pollutes the air, poisons the rivers, and decimates the forests for personal profit. At some point humans must learn to act responsibly toward the rest of creation. Otherwise, we will exhaust our resources and eventually suffer the same fate as the thousands of species we will drive to extinction by the year 2000.

The Horned Lizard: The Strangest Lizard in Texas

Texas horned lizard
Phrynosoma cornutum
Mountain short-horned lizard
Phrynosoma douglassi hernandesi
Bleached horned lizard
Phrynosoma modesrum

In 1897, workers constructing a new courthouse in Eastland County carefully sealed Old Rip in the cornerstone. The courthouse and its unwilling tenant remained in place for 31 years before a demolition crew began tearing down the old building. A curious crowd gathered as the workers reached the cornerstone. What had happened to Old Rip during the intervening years? Nobody took bets that he would be alive. Yet when workers removed the final stone, Old Rip looked up and blinked at the first sunlight he had seen in more than three decades. To the astonishment of the unbelieving audience, the bizarre lizard scampered away! Presumably it had hibernated all those years within the cool confines of its rocky enclosure.

Old Rip became the first horned lizard in the world to become nationally famous. Of course, many doubters considered the lizard a hoax, but witnesses testified to the authenticity of the event. Unfortunately Old Rip's fame and publicity were short lived. The illustrious lizard died within a year of being liberated from his stone prison. The remains of this celebrated reptile lie in state in a glass-fronted coffin in the Eastland County courthouse. Public viewing is welcomed.

Without a doubt, horned lizards are one of the most unusual animals to scamper beneath our feet. Their heads sport a crown of horns, and rows of prickly spines line their sides. The thorny growth covers the entire upper surface of their prehistoric-looking bodies. Instead of being long and slender like most lizards, they have squat, saucer-shaped bodies. Confused by their flat body, many people call these strange-looking creatures horny toads. The confusion is particularly acute in Fort Worth, where students of Texas Christian University adopted the animal as their mascot but insist on calling it a horned frog.

Horned lizards live in a world far removed from the aquatic habitat of frogs and toads. They like hot and dry places, which makes most of Texas ideal. Three species live in the state, mostly west of Houston and Dallas. At one time these novelties were common, but today they have disappeared from much of their range. I've talked to old-timers who said they once could buy them for a nickel apiece and use them for fish bait, although I can't imagine a fish striking such a ball of thorns. Some suffered from the curiosity of inquisitive children who captured them and kept them in a box. Usually the lizards outlived the interest of their young captors, who eventually returned them to the wild.

Before laws afforded them protection in 1967, pet dealers and pesticides posed the greatest threat to these slow-moving miniature dinosaurs. The widespread use of poisons killed their primary food source, ants, and animal merchants scooped them up by the bushel to ship all over the country. The transported lizards suffered the same fate as Old Rip once he was placed on display. Horned lizards cannot live very long in the temperatures we humans consider comfortable. If you kept your house above 90 degrees F, horned lizards would feel right at home.

Horned lizards, like all reptiles, depend on the sun to warm them sufficiently so their metabolic processes will work. Until they soak up enough heat, these sluggish creatures cannot convert food to energy for muscular action. Even if they are force fed, the chemistry of their digestive system does not function. At room temperature, horned lizards languish away, too warm to hibernate and too cold to eat.

Each species of animal, or plant for that matter, has an optimum set of environmental conditions under which it survives the best. The creature's behavior, color, size, and diet influence its ability to survive in its particular habitat. The optimum conditions for an animal are like trumps to a bridge player; more trumps in a hand, or habitat, mean a better chance to win, or survive and produce the next generation.

Ants are the ace of environmental trumps for horned lizards. Few other animals dine on ants, probably because they have so little nutritional content. So horned lizards have practically no competition for food, which is like having the only key to the supermarket. Specializing in ants has influenced every aspect of the horned lizard's life. It must eat vast quantities of the little morsels to get enough calories and protein to survive. To hold a sufficient meal requires a large stomach, so large that its body must be round instead of tube shaped. Horned lizards have the largest stomach of any lizard living in a similar environment.

Catching ants does not require rapid movements, a fortunate coincidence since the lizard's tanklike body design could never win a race. But sacrificing speed means sacrificing one of the best defenses most lizards have for escaping enemies. These cumbersome lizards had to develop an alternate defense. Stickers provide ample protection for cacti, so why not lizards too? Their thorn-covered bodies protect the lizards from most snakes, birds, and other predators.

When threatened, these little pincushions exhibit a temperament befitting their ferocious appearance. They arch their back, jump forward, and hiss. If the bluff fails, and the predator grabs the little animal, it discovers that horned lizards know how to use their barbed headgear. The lizard twists and jerks its head around to impale its horns in the foe and, when it can, bites and holds tightly.

Besides a studded coat of armor, horned lizards use the most successful defense in the animal world: camouflage. They can lighten and darken their colors and dappled pattern to match the soil and rocks of their environment. Cryptic coloration hides an animal only if the creature remains still, because predators easily detect movement. Horned lizards lie so flat that they cast no shadow and seem to disappear in the sand. They refuse to move until almost stepped on, then they scamper off to hide again.

If camouflage is unsuccessful and sharp horns fail to discourage aggressors, horned lizards have an unconventional weapon in reserve. When the lizards become alarmed, the blood pressure rises in their heads. Sometimes the pressure builds up enough to rupture the capillaries in the corners of the animals' eyes, and a fine stream of blood squirts out for several feet. The lizards suffer no ill effects from the episode, and the spray often deters the predators.

Lying in the sun all day eating ants or hiding in the sand would spell doom for most lizards, but the remarkable horned lizards can withstand internal temperatures that would fry most reptiles. They have the widest tolerance for high body temperatures of any lizards living in a similar environment. The ability to remain active over a broad range of temperatures gives them time to find ant beds, which may be widely scattered throughout the area.

The horned lizard's specialized diet affects its body size, defensive strategy, and even its reproduction. While most lizards lay only a few eggs, the broad body of a horned lizard enables it to carry from 25 to 35 eggs. Such a mass would slow down fleet-footed lizards, making them vulnerable to predators; however, the added weight presents no problems to slow-moving horned lizards.

The forces of natural selection have dealt the horned lizard a hand that is hard to beat. Its unique set of adaptations enables it to survive in specialized habitats in torrid desert basins and on 9000-foot mountain slopes. The 27 species in the genus live from Guatemala to Canada, from California to Texas. Whenever you see a horned lizard, remember that these little reptiles represent the versatile and creative power in nature.

The Rattlesnake: The Rattle of Danger

Western diamondback
Crotalus atrox
Prairie rattlesnake
C. viridis viridis
Timber rattlesnake
C. horridus
Northern black-tailed rattlesnake
C. molossus molossus
Mojave rattlesnake
C. scutulatus scutulatus
Mottled rock rattlesnake
C. lepidus lepidus
Banded rock rattlesnake
C. lepidus klauberi
Western massasauga
S. catenatus tergeminus
Desert massasauga
S. catenatus edwardsii
Western pigmy rattlesnake
S. miliarius streckeri

Once, as a child traveling through West Texas, I was captivated by a stuffed rattlesnake at a roadside service station. Coiled and ready to strike, the snake looked just like the pictures I had seen of the deadly serpents. I inched closer to investigate until my nose was within a foot of the angry-looking critter. Suddenly, the snake came to life with the loudest rattling I had ever heard. I fell backward in surprise and terror, amid peals of laughter. To the delight of the locals who frequented the store, the owner had pushed a button activating a raucous buzzer under the snake. The trick, probably repeated several times daily, was undoubtedly the most exciting thing happening in the entire county.

No other animal in Texas inspires as much fear, hate, and tall tales as the rattlesnake. The dread of snakebite is as much a part of growing up as sandstorms in West Texas, mesquite thorns in South Texas, and mosquitoes in East Texas. Texans have ten species of rattlesnakes to avoid, from the ill-tempered western diamondback to the diminutive pigmy rattler. The diamondback claims the western two-thirds of the state as its own, sharing portions with the Mojave, black-tailed, mottled, and banded rock rattlers, and the western and desert massasaugas. Timber and pigmy rattlers inhabit East Texas and prairie rattlers the Panhandle. Only a few counties in extreme East Texas and the Panhandle and some heavily farmed counties have no rattlesnakes.

The size of the western diamondback and the amount of venom it can deliver make it one of the most dangerous snakes in the world. Fortunately for humans, a majority of rattlers would rather flee than fight, attacking only when escape seems impossible. They depend on retreat and concealment by camouflage as their best defenses. If discovered, their rattle warns the unwary to move away. This strategy worked well with bison and other large herbivores, but in modern times it has put the snake at a disadvantage. The taming of the West brought with it the universally accepted law of killing all rattlers, a general rule usually interpreted to include all snakes. Rattlesnakes less inclined to rattle seem to be more adapted for surviving in today's world.

Rattlesnakes belong to the pit viper family, a group of snakes including copperheads and water moccasins that have a unique set of sense organs. These snakes have pits on either side of their nostrils containing a high concentration of extremely sensitive heat receptors. The ability to "see" the body heat of warm-blooded creatures enables pit vipers to hunt effectively at night. Like a heat-sensing missile, they seek out their prey and strike with deadly accuracy.

A snake has no ears or ear openings, a benefit to an animal that crawls and burrows in the ground. Although deaf, these creatures can sense the ground vibrations of approaching large animals, such as humans. Keen vision is another sense sacrificed by a snake's ancestral burrowing lifestyle. With hearing and sight of little value to a snake, smell has become its most vital sense. Scent is so important that snakes, and some lizards, have a pair of extremely sensitive sensory organs in the roofs of their mouths to help them find the live food that they eat. A snake flicks out its forked tongue to pick up airborne chemical molecules and transfer them to the receptors, called Jacobson's organs. Since their other senses are so limited, snakes use these organs as their primary contact with the world, as evidenced by their constant tongue flicking. A male snake also uses its nose to follow the pheromone, or scent, trail of a female ready to mate, and a poisonous snake uses smell to track prey that scamper a few yards away before dying.

Unlike nonpoisonous snakes that depend on speed to catch their prey, the heavier-bodied rattlers lie in wait and use venom to kill the rodents and other small mammals they eat. After killing the animal, the primary function of the venom is to begin the digestive process from the inside out. Most of the toxic components contribute to breaking down tissue and muscle, allowing the snake to assimilate the food rapidly. In human poisoning, the digestive toxins may cause serious internal hemorrhaging and swelling. With deep fang penetration, extensive muscle damage can result in impairment or loss of limb.

Snake venom is a complex fluid containing 12 to 30 peptides and enzymes. Some destroy blood, others begin digesting tissue, and others attack the nervous system. Some components attack particular organs, and others break down tissue to increase the effectiveness of other components. The Mojave rattler, with more neurotoxin in its venom than the diamondback, has the most deadly venom of any rattler in the state. Venom immediately combines metabolically with the surrounding tissue, making removal by suction impossible. Tourniquets, which have caused more injury and loss of limbs than snakebites, prevent the spread of toxins, but without life-giving blood, the cells in a limb will die in 30 to 40 minutes. Once a limb dies, it must be amputated.

Rattlesnakes play an important, though often overlooked, role in controlling rodents and rabbits on the range. As the rodent population increases, the numbers of snakes and other rodent-eating predators increase proportionally. Without the predators to keep the rodents in check, rats, mice, and rabbits would soon overrun the range. The government spends millions of dollars annually in Texas fighting rodents, which inflict heavy economic losses on ranchers and farmers, yet nature provides the best and least expensive form of vermin control.

As night falls and the summer heat abates, rattlesnakes emerge from hiding and begin their search for food. When hunting, rattlesnakes do not forage randomly throughout their range. An Oklahoma study that outfitted 83 rattlers with radio trackers discovered they prefer selected routes that provide protective cover. In the hottest part of the summer, they may forage for only a few hours around midnight and then retreat to a protected den for the day. They sometimes hunt during daylight in the spring or after a cooling summer rainstorm, but usually remain nocturnal. In the winter, Texas rattlers do not truly hibernate and may emerge from their dens on sunny days to bask. In South Texas, they remain active during most of the year.

South Texas diamondback rattlers feed well into the winter months, enabling the females to reproduce every year. Northern rattlesnake species require a year of storing fat in order to produce a litter. They do not have enough time between parturition and hibernation to accumulate enough reserves to reproduce the next year. The 9 to 16 young, born alive usually during September, average 9 to 13 inches in length depending on the species. The babies emerge with a button and a lethal load of venom. They shed and get another rattle in about one week. A snake adds a rattle each time it sheds, about every six to eight weeks when food is plentiful. During an eight-month feeding season, a rattler can add four to five rattles, but long strings are unusual because the rattles break off easily in the brush and rocks.

The western diamondback holds the record for size, both in length and in the fictitious tales it has inspired. Texas author and historian J. Frank Dobie compiled an entire book of rattlesnake folklore. He described a specimen said to be about 15 feet long that once decorated a rail car running between Brownsville and St. Louis. Since a skin will stretch as much as 33 percent, the snake must have been about 12 feet long, if really a rattler. Stuffed pythons have been painted convincingly and, with rattles attached, passed off as giants. W. S. "Snake" King of Brownsville, well known for both his king-size snakes and stories, claims to have captured a 9-foot, 6½-inch diamondback. Seven-foot individuals are well documented, which is probably as large as diamondbacks grow. The majority of adults grow to 3 or 4 feet, regardless of age, yet populations of snakes, like the human population, include a few genetic giants. King's offer of $100 per inch for any western diamondback over 6 feet went unclaimed for years.

Most stories about rattlesnakes involve giant size, snakebite, or folk remedies. According to legend, a rattlesnake bite is so poisonous that it will kill a tree if the snake accidentally misses its victim and strikes the plant. To prove its point, the story relates the time a rattler bit a wagon tongue. The tongue began swelling, and the driver had to cut it off to save the wagon. A mythological snake once bit a tire, and the high-pressure air rushed through its hollow fangs, filling the unfortunate creature like a balloon until it exploded.

One of the most persistent fables, dutifully passed on to me in my youth, concerns a beautiful woman and her fiancé. On their wedding day, he bought new boots and on the way to the church received a fatal bite on the foot. After the funeral, his best friend began courting the bride-to-be, and they eventually announced their plans for marriage. Since his late friend's boots fit him, he decided to wear them. During the ceremony, he suddenly dropped dead. A third friend consoled the grief-stricken woman, and as time passed, they too fell in love. On the day of the wedding, he decided to wear the ill-fated boots. While putting them on, he noticed a broken snake fang waiting to inject its lethal venom into the unlucky person who inserted his foot.

Every snakebite story can be matched by the tale of a sure-cure remedy. Many involve an elusive herb called rattlesnake weed. None of the stories ever really identify this mysterious plant that when chewed, applied as a poultice, or made into a tea cures snakebites, but numerous plants bear the common name. Other alleged cures, besides the traditional bottle of whiskey, include jabbing the wound with yucca or agave spines and using fresh liver or meat to draw out the poison.

The apparent success of the folk remedies depended on how much venom, if any at all, the snake actually injected into the victim. One-fifth of the time rattlesnakes deliver no poison, and about 70 percent of the bites inject venom only superficially. Doctors today advise that the standard cut-and-suck treatment taught to every Boy Scout should be discarded with the whiskey and rattlesnake weed remedies. No treatment at all is better than cutting, tourniquets, or packing with ice for long periods. The recommended first aid is to wrap an elastic bandage around the limb between the bite and the heart. Secure the bandage snugly enough to compress the lymph system but not restrict the flow of blood. Immobilize the affected limb in a splint, remove shoes or rings in case of swelling, and rush to a hospital.

Today rattlesnakes continue to be as much a part of modern Texas culture as in times past. Each spring Sweetwater, Freer, Taylor, and other towns sponsor carnival-like rattlesnake roundups. To the cheers of excited masses, contestants see who can bag the most snakes without getting bitten, stay the longest in a sleeping bag filled with rattlers, and take part in other events exhibiting equally absurd bravado. Visitors can dine on rattlesnake burgers and buy snakeskin hat bands and belts, rattler key chains, and other oddities. Many animal lovers protest the debauchery and inhumane treatment of the reptiles, but rattlesnakes find few advocates in the ranks of Texas lawmakers. Perhaps someday we humans will learn to value all life, and not treat other creatures as objects for our entertainment.

Turtles: The Animals with a Mobile Home

Modern vacationers who set out in their recreational vehicles each summer have adopted a lifestyle turtles perfected 200 million years ago. Like the turtle, today's travelers live in safety and comfort by taking their homes with them. Through the eons, this strategy has proved successful for the turtle. Dinosaurs, the largest and fiercest reptiles, vanished from the changing world, yet the unaggressive turtle lives on. This small reptile, with its home on its back, crawled out of the golden age of dinosaurs, across the drifting continents, and into the present age of mammals.

While many people have a hysterical fear of snakes and find lizards repulsive, turtles capture the imagination of us all. From children to adults, people all over the world like them. Pottery, paintings, and myths show that even ancient cultures favored turtles. Today art, children's stories, and sage expressions glorify and anthropomorphize this creature's characteristics. "Observe the turtle," the saying advises, "it doesn't get anywhere unless it sticks out its neck." Remember the tortoise and the hare? We may laugh at the turtle's slowness, but we admire its persistence.

Baby turtles often become children's first pets, their introduction to the world of nature. A quarter-size red-eared slider, *Trachemys scripta elegans,* paddled around in my first aquarium. Later I found an ornate box turtle, *Terrapene ornata ornata,* and fed it moths attracted to our front porch light. After discovering how hard these toothless critters could bite, I released it, but not before painting my phone number, LO-2975, on its back. The next day, a friendly neighbor three blocks away called for me to retrieve the wandering turtle. The distance that poky little creature traveled left a lasting impression on its young captor. I learned later that painting a turtle is like painting your skin, an unhealthy thing to do, and that most pet turtles die from the inadequate diet of commercial turtle food.

The ancestors of modern turtles saw the ice ages come and go, mountain ranges build and wear away, continents collide and reform, catastrophic meteor showers, and massive plant and animal extinctions. To survive these extremes, turtles had to be exceedingly resilient to environmental changes. These well-adapted creatures even can withstand radiation levels that would kill most animals. Perhaps turtles, along with cockroaches, will eventually inherit the earth.

In surviving the millennia, turtles have solved three of an animal's biggest problems: food, protection, and shelter. Some animals specialize in a particular type of food, but most turtles eat anything, plant or animal, that they can grasp in their horny jaws. Their varied diet kept them well fed while the plant and animal community around them continuously changed through the ages.

Many animals spend most of their waking hours trying to avoid becoming another animal's meal, but turtles carry their protection on their backs. Their bony armor both protects them from hungry predators and provides an instant home. The turtle's shell is its most remarkable adaptation for survival, and separates these animals from all other creatures. A turtle's ribs and vertebrae are fused into its upper shell, or carapace, placing its hip bone inside the rib cage, a unique arrangement among vertebrates. A bony bridge on each side connects the carapace to the bottom shell, the plastron. Living encased in a horny bunker offers protection, but makes locomotion cumbersome. Many, but by no means all, turtles circumvent this problem by taking to the water, a much more buoyant medium than air. But even water turtles return to land to bury their leathery eggs. The female digs a shallow hole and lets the soil, heated by the sun, incubate the eggs.

To illustrate their incredible ability to adapt to extreme environments, some turtles, instead of swimming through lakes and rivers, trudge across the sand in the hottest, driest deserts in the world. The desert tortoise of the Mojave Desert and the Texas tortoise, *Gopherus berlandieri,* both relatives of the giant tortoises of the Galapagos Islands, lumber through cactus-covered landscapes eating grass, stems, leaves, and prickly pear pads and fruit.

Turtles and other reptiles have another basic survival strategy that many animals find limiting: their body temperature rises and falls with the external temperature. Mammals and birds expend vast amounts of food energy maintaining a constant temperature, while reptiles merely climb on a rock or limb and let the sun do the work. Once their body reaches a temperature warm enough for the chemistry of muscle action and digestion to function, reptiles become as agile and fleet as mammals. These creatures can survive on a tenth of the food energy required by a mammal, a reasonable trade-off for the ability to function at low temperatures.

Turtles are particularly abundant in North America. Of the approximately 200 species in the world, about 86 live in the United States, with some 35 calling Texas home. In contrast, 6 species live in Europe, 19 in Australia, and 35 each in Africa and South America. Most turtles live in fresh water, and many of the larger river basins in the nation have their own unique species and subspecies.

The alligator snapping turtle, *Macroclemys temminckii,* occurring from East Texas to West Florida and along the Mississippi River, holds the size record for a freshwater turtle. These mammoths of the lakes and rivers sometimes weigh more than 200 pounds. They have powerful jaws to crush their prey and have been known to break the fingers of unwary humans. To trick fish to venture within striking range, alligator snappers have a wormlike lure on the inside of their mouth.

Perhaps the strangest turtles of all ride the ocean

waves. The seven species of sea turtles in the world have adapted well to their watery existence. Their front feet are elongated flippers that propel them along as fast as 20 mph. The leatherback turtle, *Dermochelys coriacea,* grows to be the largest living reptile. In 1923, fishermen captured an 8½-foot, 1286-pound leatherback in Monterey Bay, California. Unlike other turtles, the leatherback, as its name implies, has no horny carapace but many small bones embedded in a thick, leathery skin.

Except for brief trips to selected beaches by the females to lay eggs, sea turtles spend their lives swimming the tropical and subtropical seas. These creatures have mystified sailors and scientists for centuries. Marine biologists have watched the females laboriously digging their nest chambers on isolated beaches and laying 100 to 150 eggs each, seen the hatchlings scampering toward the surf, and occasionally captured an adult swimming at sea. But where the juveniles spend their lives remains for the most part a mystery. Despite years of research, biologists know little about what happens between hatching and maturity.

The Kemp's ridley sea turtle, *Lepidochelys kempi,* a denizen of the Gulf of Mexico and Atlantic seaboard, puzzled scientists the longest. For 18 years, marine biologists searched in vain for the breeding sites of this mysterious beast. Some people considered it a hybrid between the loggerhead and green sea turtles, since a female with eggs had never been found. Finally in 1961, a film made in the 1940s stunned the audience at an American Society of Ichthyologists and Herpetologists meeting in Austin. A Mexican had filmed approximately 40,000 ridleys nesting in one day on a 15-mile stretch of beach near Tampico.

Villagers long had known what had eluded the biologists, and when the turtles massed offshore they waited with bags, boxes, and trucks to gather the eggs. The eggs, considered an aphrodisiac, brought handsome prices as far away as Mexico City. By the time the scientific community discovered the nesting beach, less than a thousand turtles remained and the species was on the brink of extinction. In 1966, the Mexican government finally legislated protection and sent soldiers to patrol the tiny stretch of beach that is the birthplace of every Kemp's ridley sea turtle.

Outdoorsman Dearl Adams and his wife, Ethel, of Brownsville realized that the ridley needed more than conservation to save it from extinction. A second nesting colony would double the magnificent turtle's chance of survival. Between 1963 and 1967, they transferred 5000 eggs from Mexico to South Padre Island, which closely resembles the turtles' native beach. Assisted by Ila Loetscher, who became nationally known as the Turtle Lady of Padre Island, they lived on the beach each year to protect the eggs and hatchlings until the tiny turtles swam away in the rolling surf. Finally, in 1976, their dedication was rewarded. A 100-pound ridley swam ashore and laid 97 eggs.

The U.S. government realized the value of the project, and in 1978 began a large-scale operation to transfer eggs to Padre Island National Seashore. Each year, scientists bag 2000 eggs, pack them in Padre Island sand, and take them to the seashore to hatch. The hatchlings are given a brief taste of freedom in the sea, and then netted and taken to facilities in Galveston for a year. The scientists hope that the exposure to the sand and sea of Padre Island will imprint the new location in the babies, and they will return there as adults to nest.

In the wild, only about 1 percent of the baby sea turtles reach maturity. The rest become meals for everything from crabs and gulls to fish and sharks. By rearing the young in a protected environment for the first year, about 30 percent are expected to reach breeding age. Workers tag and release the eight-inch yearlings about 15 miles at sea.

The life history of the ridley still remains a mystery. Except for a few washed across the Atlantic by strong currents, adults occur only in the Gulf of Mexico. Juveniles have never been recovered in the Gulf, only in bays on the Atlantic coast. Perhaps the young leave the Gulf for a period and return as adults.

The future of the Kemp's ridley sea turtle remains uncertain. The Padre Island transplants may take 10 to 15 years to mature sexually. In 1986, 8 years after the start of the program, rangers began patrolling the beach regularly looking for nests, but only rare, sporadic nesting has occurred on the island. In the meantime, fewer females nest in Mexico each year. The number of eggs decreased from about 100,000 in 1978 to 45,000 in 1986. To compound the problem, shrimpers' nets increasingly threaten the few turtles that escape predators and reach maturity. The industry has fought the use of turtle-excluding devices that would keep the air-breathing animals from drowning in the nets. Apparently, only legislation will force shrimpers to use the lifesaving device.

After millions of years of swimming freely in the open sea, the ridley sea turtle now depends for survival not on its natural adaptations but on the cooperation of the Mexican, United States, and Texas governments, conservation groups, and the fishing industry. Time will tell if this mysterious creature of the deep will survive the pressures of the twentieth century.

Invertebrates

If an extraterrestrial were cataloguing life on this planet, invertebrates would lead the list in numbers. The invertebrates include all animals without backbones, from sponges to the ubiquitous insects, from tiny soil nematodes to the giant octopus. Only a few classes of animals with a centralized brain and a complex neural system have backbones, and these are far outnumbered by the invertebrates.

The land, sea, and soil teem with invertebrates. The minuscule zooplankton of the sea are the foundation of the food pyramid that supports all ocean life. Larger invertebrates, like shrimp and clams, and many fish, filter the tiny organisms from the water, and in turn become food for larger fish. Without the abundant invertebrates, life in the sea would collapse. Insects play a pivotal role in the terrestrial ecosystem. Without insect pollinators, many plants, and all the animals that depended on them, would disappear. Without the soil organisms decomposing the dead animals and plants, the nutrient balance would be disrupted. Without the insects and other invertebrates providing the basic fodder for the land animals, the food chain would shatter.

In comparison with the invertebrates, the more highly developed forms of life play an almost insignificant role in the biosphere. If the mammals, birds, and reptiles suddenly disappeared, life on the planet would proceed with little change, but removing the invertebrates would totally disrupt the fabric of life on the earth. Vertebrates may be the star performers in the pageant of life, but the play would fold overnight without the supporting role of the invertebrates.

Scientists have divided all life into five distinct king

doms. The Monera kingdom contains bacteria, and probably will be divided into two separate kingdoms. The cells of these organisms have no nucleus and, in terms of evolutionary development, represent the simplest forms of life. The Protista kingdom consists of predominantly single-celled organisms like the amoeba and algae. Scientists undoubtedly will divide this kingdom when the origins of its diverse members are better understood. Molds, mildews, mushrooms, and yeasts belong to the fungi kingdom. No one knows where to place the viruses, but they probably will have a separate kingdom of their own. The rest of life fits into the two more familiar, but vastly complex, plant and animal kingdoms.

Biologists have subdivided the animal kingdom into nine major categories according to the complexity of cellular, tissue, organ, and body development. All are invertebrates except for the few classes of animals with backbones. Sponges have a body composed of a few specialized cells largely independent of each other. Corals and jellyfish, including the Portuguese man-of-war described in a following section, have cells organized into tissues with only a few rudimentary organs, such as tentacles. Animals with major organs are divided into those with no body cavity, the flatworms, and those with a fluid-filled body cavity. Major divisions within the latter group include the earthworms, nematodes, mollusks, arthropods, sea urchins, and chordates. The joint-leg arthropods, the most diverse group of living animals, include crabs, shrimp, spiders, mites, and insects. The chordates include the vertebrates, composed of the fishes, amphibians, reptiles, birds, and mammals, as well as some invertebrates.

If our extraterrestrial explorer were choosing the most successful land animals to use to colonize a distant planet, the choice undoubtedly would be the insects. Terrestrial life is largely a celebration of this incredibly diverse class of invertebrates. Insects outnumber all other plants and animals in both species and individuals. The six-legged critters comprise about 80 percent of the animals on the earth. Scientists have described nearly one million species, and as many as one million more may wait their turn to be discovered.

Insects became the most numerous, widespread, and diverse forms of life because of their ability to adopt extremely specialized lifestyles. By dividing the food resources into narrow niches, numerous species can exist side by side without competing with each other. An insect's mouth, with either chewing or sucking parts, and its digestive system adapt it to a particular food source, often only one narrow group or species of plants or animals. Mosquitoes have mouths designed for sucking blood, most butterfly caterpillars feed on a single type of plant, and scarab beetles specialize in eating manure. A leaf miner burrows through the inside of a leaf, a gall wasp lays its eggs in the leaf, twig, or branch of a single plant species, and the tarantula wasp attacks only spiders. Insects and plants have evolved a complex dependence on each other: showy, fragrant blooms attract insects, which pollinate the flower and receive a nutritious reward for their efforts. In contrast, a plant that depends on the wind to spread its pollen has tiny, nondescript flowers.

Having a narrow food preference enables insects to diversify, and their robust ability to survive adverse conditions allows them to exploit every imaginable habitat. They live in glacial snowfields, arid deserts, deep lakes, and on stalactites in caves. In California, a fly larva lives in pools of crude oil in petroleum fields, and another eats mold in vats of corrosive formaldehyde. Insects even live in such extraordinary microhabitats as the nostrils of reindeer and the water in hollow leaves.

After 300 million years of adapting, insects have the odds of survival in their favor. Most insects cover their bets with a simple strategy of copious reproduction. If each female produces thousands of offspring, some invariably will be able to survive almost any adverse condition. We spray them with poisons, but succeed only in eliminating the susceptible ones. Each generation will have genetically resistant individuals that live to produce more poison-tolerant young. Many of today's insects can walk through DDT with impunity. One species of grasshopper even concentrated a particular insecticide in its body. If the poison deters other animals from eating the grasshopper, we have increased its ability to survive instead of killing it. Insects could have written the book on natural selection and survival of the fittest.

Except for the few creatures that have an impact on our daily lives, most invertebrates live in a complex world that we normally overlook. Ants and termites have an intricate social system that governs colonies as populous as our large cities. Bees communicate the location of nectar-rich flowers by an elaborate dance. Insects can see spectra of light beyond the sensitivity of the human eye and hear frequencies beyond the audio range of our ear. A Saturniid moth has such a keen sense of smell that the male can detect a female as far as five miles away. Insects developed flight 100 million years before birds soared through the sky, which helped them spread around the globe and diversify into the millions of living species. Insects vary in size from the 1/100-inch hairy winged beetle to several moths with a 12-inch wingspan.

Invertebrates greatly influence our daily life, both positively and negatively. We may curse the discomfort caused by blood-sucking mosquitoes, pesky flies, and stinging ants, wasps, and scorpions, but we thrill at a dancing butterfly, savor lobster, shrimp, and oysters as culinary delicacies, enjoy our silk clothes, and consider honey a staple food. Insects pollinate the vast majority of the plants, provide a source of food, either directly or indirectly, for most of the other animals, and complete the food cycle by decomposing dead plants and animals into nutrients that plants can assimilate. Just one acre of forest soil contains about 65 million soil insects. Insects and insect products, such as beeswax, provide medicines, dyes, shellac, lubricants, salves, ointments, wax, polish, and varnishes. Although whole industries have developed to help control insect damage to agriculture, most species benefit the farmer and rancher by controlling weeds and other plant pests and aerating and fertilizing the soil. Even though insects spread disease and destroy food, timber, and other necessities worth $3.5 billion annually in the United States, these animals directly benefit the economy by $4.5 billion, a net gain of $1 billion.

The next six sections describe some of the hidden mysteries of the invertebrate creatures that we commonly encounter in our backyards, at the beach, and in the woods. Read about a colony of organisms that float on the ocean waves, a fragile butterfly that migrates thousands of miles, and critters that sting, itch, and suck our blood.

Chiggers and Ticks: Pint-size Pests

Chigger
Trombicula alfreddugesi
American dog tick
Dermacentor variabilis
Lone Star tick
Amblyomma americanum

The humans, oblivious of the blood-hungry predators, marched merrily through the dense vegetation. Hidden in the foliage, the carnivorous beasties waited with claws outstretched, ready to snare their victims, tear into their bodies, and gorge themselves on the juicy flesh and blood. This may sound like the opening scene of a horror movie, but it is only a typical spring picnic. While we play and cavort in the park, chiggers and ticks hang on the tips of grass and leaves with outstretched legs waving in the air, waiting to grasp any animal that passes by. Once they have hitched a ride, they search for a tender patch of skin for their own picnic. For humans, these parasites cause irritation, infection, and, at worst, disease.

Chiggers and ticks both belong to the class Arachnida, which also includes scorpions and spiders. Each develops in a completely different way, and each poses different problems for humans. While chiggers mainly cause physical discomfort, ticks can carry the potentially deadly Rocky Mountain spotted fever (RMSF).

Texans are lucky. Of the more than 2000 species of chiggers in the world, only 2 bother people in our state. Some chiggers in other parts of the world specialize in a particular animal, but ours dine on a large number of hosts. Besides human flesh, they feast on mammals, birds, reptiles, and even amphibians.

Chiggers pass through four life stages: egg, larva,

nymph, and adult. The nymphs and adults, only 1/20 of an inch long, perform a service to humans by eating insect eggs, small insects, and other tiny organisms. The pin-size larvae, 1/150-inch, are the tiny culprits that inflict king-size misery in humans. These minuscule pests invariably homestead on the tenderest parts of our bodies, especially waists, armpits, groins, and behind the knees.

The larva works its way into a skin pore or hair follicle and attaches with two pairs of barbed mouth appendages. Then begins a scene that surpasses any monster movie ever produced by Hollywood or Japan. The chigger slowly begins to dissolve away the body of its victim. Fortunately a microscopic parasite can only liquefy microscopic amounts of flesh. The digestive enzymes the chigger pumps in form a hollow strawlike tube that the critter uses to lap up the dissolved skin cells. The larva gluttonizes for three or four days and, when engorged, drops off to develop into the nymph stage.

Chigger bites itch so intensely because humans are not the perfect hosts. Our body recognizes the pest as an intruder and reacts strongly. A puffed-up red bump forms around the parasite and usually persists for several days after the villainous arachnid departs. Once you become a chigger landlord, you will probably have to live with the unwelcome tenants until they leave voluntarily. Tape may pull them off, and medicines may kill them, but the itch remains. I know from experience that the itching from several hundred bites will almost drive you insane. The three or four days of constant misery seem like an eternity. Some chiggers in Asia and Australia are adapted to humans and cause no itching, but they spread deadly diseases.

The best way to survive chigger season, spring through late summer, is to avoid the pests as much as possible. Spray insect repellent on your ankles and pants legs, and scrub with soapy water after being outdoors. Chiggers hate sulfur, and dusting your ankles with an old sock filled with the yellow powder works as well as expensive sprays. Some people even dust their yards, which may deter the pests until the next rain. Most drugstores sell powdered sulfur.

Ticks are another pint-size parasite eager to exploit the protein available in the human body. Using specialized mouth parts, a tick buries its head in the victim's skin and, like a tiny vampire, begins sucking blood. Ticks feed on the blood of animals during every stage of their life except as eggs. A female tick lays 3000 to 6000 eggs in the spring, which hatch into tiny "seed" ticks. The larval seed ticks begin the lifelong process of grasping for any animal that passes by. They engorge, molt, and grow into nymphs, then finally into reproducing adults. At any state of development they can transmit diseases to humans.

Ticks carry Rocky Mountain spotted fever and, as vectors of the disease, pass it to other animals. The malady, also called tick-borne typhus fever, occurs throughout the United States and south to Brazil. Ticks become infected by feeding on diseased animals and carry the pathogen for life, about 18 months. Females transmit the fever to their many offspring. Prime tick season in Texas and the rest of the United States occurs from April through October. During this season, anyone who spends time in wooded, bushy areas should inspect for ticks daily.

The incidence of the fever increased from 433 cases nationwide in 1969 to 1165 in 1981, but the number of cases fluctuates with peak and low years. Texas recorded 108 cases in 1983, yet only 33 in 1985. The fatality rate averages about 5 percent. Hunters and campers contract the disease much less than those who live in tick-infested areas. Most cases occur in wooded East Texas among rural inhabitants. Children under 15, especially between 5 and 9, contract RMSF more than any other age group, probably because of continuous exposure to the outdoors and from ticks on pet dogs.

Only about 1 percent of the ticks carry the RMSF microorganisms. Symptoms of the disease develop within three to ten days, beginning with a measleslike rash on the hands and feet that spreads over the entire body. Severe muscle aches and headaches, chills, and fever accompany the rash. The general nature of the symptoms and the fact that most doctors have no experience diagnosing the disease can have serious consequences. Without treatment, the symptoms last for two to three weeks with a death rate of about 20 percent. When treated by the first or second day of the rash, the fever usually falls within 48 hours. With the cure the victim receives a bonus: lifelong immunity.

Of the many species of ticks, only two carry the disease in Texas, the American dog tick (*Dermacentor variabilis*) and the Lone Star tick (*Amblyomma americanum*), named for the white spot on its back. In the West, the Rocky Mountain wood tick (*Dermacentor andersoni*) also transmits the disease. To spread Rocky Mountain spotted fever, a tick must be attached for two to six hours. The Federal Centers for Disease Control advise against using a hot match, petroleum jelly, or any other method that kills the tick while it is embedded. Injuring the tick could cause it to discharge its fluids, including the disease organisms, directly into your body. The crushed tissue or body fluids can also transmit the fever. For this reason, never mash a tick between your fingers. You can safely remove a tick with tweezers or tissue-covered fingers. Grasp the tick by the head as close to the skin as possible and roll the body upward without jerking or twisting. Ideally the tick comes out alive with a bit of skin still in its mouth. Thoroughly wash and disinfect the bite and fingers.

You can greatly decrease the chance of a tick bite by tucking in shirttails and pants legs. Spray with a commercial insect repellent before heading into the woods or backcountry. Closely check clothing for loose ticks. The critters may crawl around for several hours before attaching. Also, brush long hair and check your scalp. If you develop a rash or fever within ten days of exposure to tick country (only 80 percent of fever victims recall an actual bite), consult your physician immediately.

The Fire Ant: Wildfire with Six Legs

Imported red fire ant
Solenopsis invicta

They swarm, they sting, they attack with a fury. Just as locusts smote the Pharaohs of Egypt and plague-ridden fleas decimated Europe, the imported red fire ant has assaulted the United States. Undetected, the $^1/_8$- to $^1/_4$-inch ants hitched a ride on a Brazilian cargo ship sometime in the 1940s and jumped ship in Mobile, Alabama. By 1953, they had pillaged their way through the South like General Sherman and reached the Texas border. Texans may have wrested their vast state away from the Indians and Mexicans, but they could not stop the relentless advance of the fire ants. Nearly a quarter of the entire state has fallen to the invaders. Like flags of conquest, the conical mounds of these indomitable insects now decorate 113 of the state's 254 counties. Fire ants outnumber Texans ten million to one!

Fire ants are more than a pest—they represent a serious menace to ranchers, farmers, and urban dwellers. The one-to-two-foot-high mounds of a mature colony may contain 250,000 ants, which swarm out ready to attack on the slightest provocation. Even a gentle puff of air on the mound will incite the belligerent critters. Infested areas have as many as 50 mounds per acre.

The crusty mounds can render a pasture or field virtually unusable for ranchers and farmers. Cattle wisely avoid pastures infested with the aggressive ants, and farmers find the sun-hardened mounds worse than speed bumps on a freeway. The mounds damage equipment, jam combines, and dull blades. Farmers sometime leave their crops in an overrun field if harvesting would seriously damage their equipment. The ants often ruin hay by homesteading in the bales. Each year Texans spend about $1.5 million, and Americans spend $6 million, trying to control fire ants. Soil, plants, sod, hay, and logs are a few of the quarantined items that cannot be transported from infested counties into uninfested areas without stringent control measures.

Of the 10,000 species of ants worldwide, the imported fire ant poses the most serious threat in North America, especially in residential areas. They find manicured lawns, parks, school grounds, and roadsides perfect homesites. A well-watered sunny expanse of turf and the absence of other ants present the perfect invitation to these adventurous insects. Most people in the eastern half of the state have had close encounters of the stinging kind with fire ants, and many in their own yards.

Fire ants derive their name from the intense pain caused by their sting. A burning sensation begins as soon as the ant injects its venom; the skin turns red and a pustule forms. The pus irritates the surrounding tissue, causing itching. Bacteria may enter and cause secondary infections if the pimplelike sores are broken by scratching. The sores usually heal in about a week.

Angry fire ants sting until killed, and their swarming attacks occasionally hospitalize their unfortunate victims, particularly small children. A surprised highway patrolman in East Texas abruptly forgot the speedster he was ticketing when he discovered he was standing in a fire ant bed. Doctors in one county in Georgia treated almost 30,000 fire ant sting cases in one year. Once I asked my wife and two children to pose for a picture on a stump. As I took the picture, my daughter said, "Daddy, there're ants in the log!" The apprehensive expressions in the snapshot foretold the dancing and screaming that immediately followed.

When a person develops an extreme allergy to fire ant venom, a single bite can prove fatal. In 1985, two people died in Texas. One, a truck driver, laid his head in an ant bed while changing a tire. He recovered from the painful experience, but suffered a fatal heart attack several months later when three ants bit him on the toe. If the symptoms of anaphylactic shock develop—dizziness, sweating, swelling of the throat—seek immediate medical attention. A half-and-half solution of water and chlorine bleach or a paste of meat tenderizer provides some relief from the stinging, itching, and pustule formation if used immediately. A doctor at Texas A&M University found that injections of puréed fire ants help desensitize people to the venom.

Unwelcome, but unhesitating, fire ants move into any neighborhood and may even try to share your house. A New York couple that moved to Austin found them in their washing machine and the walls of their laundry room. By taking advantage of the warmth of our houses and the dampness of our irrigated lawns and crops, these tropical ants can spread beyond the natural barriers of cold and dryness. Although the dry West Texas summers and a winter monthly average low of 10 degrees should have halted their expansion into North Texas, these pests have survived some of the harshest winters in history in Lubbock, and they may eventually become scattered throughout the Panhandle.

Despite the combined efforts of federal and state agricultural departments, the invaders have not been stopped, or even slowed. Since their introduction into the United States, they have advanced steadily at a rate of 5 to 12 miles per year. From a modest beginning of perhaps only one queen, fire ants have infested 11 states, an area of more than 230 million acres, with 50 million acres in Texas. If they inadvertently are transported to the West Coast, they will spread from Arizona to Washington.

Texas has three species of native fire ants that cause few problems compared with the imported species. They are less aggressive and have smaller colonies, and natural factors control their spread. The native fire ants require three to five years to produce queens, while the imported red fire ant takes only seven months. A single colony of imported ants may produce 4500 queens per year, but fortunately for the world, 90 to 99 percent never survive to establish colonies.

Imported fire ants are prime examples of animals adapted to pioneering into new territory. Colonies of ants form when new queens mature and leave the nest. On sunny, windless days in the spring and summer, the virgin queens and drones mass at the mound surface and fly skyward to mate. After the nuptial flight, the fertilized queen lands ¼ mile to 1 mile away and loses her wings. Now earthbound, she lays 10 to 15 eggs that hatch as tiny workers in about 28 days. Gradually the number of foragers, nurses, and other caste members increases until new queens can be produced.

When foraging for food, imported red fire ants outcompete the native species. Besides being more aggressive, they win by numbers alone. A native fire ant colony contains anywhere from 100 to 5000 individuals, but there are 250,000 per colony of the imported species. Red fire ants build subways about ½ inch underground to reach their foraging sites safely. The tunnels radiate as far as 75 feet in all directions from the mound. Workers exit through openings along the way and forage at random until they find a food source. Then they return to the mound, leaving a pheromone, or scent, trail for other workers to follow.

A foraging imported red fire ant stalking through the weeds is a vicious predator armed with a potent weapon. It uses its venom to kill spiders, ticks, insects, earthworms, and other small invertebrates. It may even attack ground-nesting birds. As predators, fire ants sometimes aid farmers by controlling crop-damaging insects, such as boll weevils and corn earworms. However, the native ants they displace may perform this service as well or better. Although hunters, these omnivorous ants eat almost everything in their path. They eat plant sap, germinating seeds, and buds, and tend aphids for the honeylike secretions the aphids produce.

Until banned by the U.S. Environmental Protection Agency in 1978, Mirex, a fat-soluble compound more persistent than DDT, was the pesticide most used to control fire ants. In some areas 44 percent of the people had accumulated residues of this chlorinated hydrocarbon in their bodies. After 16 years of widespread use, Mirex proved to be carcinogenic, caused birth defects, and killed a large number of nontargeted animals.

Amdro, the pesticide now approved for broadcast application for nonagricultural land, does not accumulate in the environment and dissipates within 24 hours. It will kill 85 to 90 percent of the ants of all species when spread on rangeland, pastures, and lawns, but in 6 to 9 months fire ants will reinvade. Since the chemical degrades so fast, it must be used in the cool of the morning or evening, when the foraging workers will carry the poison back to the mound to feed the queen.

After years of research, a juvenile hormone pesticide, Logic, has been approved for areawide broadcast. Worker ants carry the pesticide into the mound and feed it to the larvae. The hormone disrupts the larvae's metamorphosis into adults, and the colony dies from lack of workers. Although formulated to kill ants, Amdro and Logic also kill aquatic organisms and should not be used around water. Research continues on other ways to control the imported fire ant biologically. In Brazil, researchers have discovered about ten different pathogens and parasites that naturally limit the fire ant population.

Areawide application of pesticides is not always the best way to control fire ants because the native ants in an area help keep the imported species in check. When all species of ants have been killed by pesticides, imported fire ants repopulate more efficiently than the native species. Within a year after broadcasting pesticides over an area, the fire ant population may have increased 3000-fold. For this reason, only individual mound treatment is recommended for low-density infestation, less than 15 mounds per acre.

A number of pesticides are approved for individual mound treatment. You can spread baits for the workers to carry back to the colony, or soak the mounds with drenches. The baits should be spread uniformly for three feet around the mound, or use at least three gallons of liquid to soak the mound. For homeowners, boiling water, which kills 60 percent of the mounds, may be the easiest and safest approach. Do not disturb the ants when treating a mound. Excited ants immediately carry the queen deep inside the mound, which may extend three feet below the surface, and a colony will survive until its queen is killed. Never use gasoline or other petroleum products because of their flammability and damage to the environment.

Each year, specialists from around the nation gather for a symposium to report the latest biological research and developments for controlling the imported fire ant. Control, not elimination, is the goal. Reagan Brown, the former Texas commissioner of agriculture, expressed the feelings of both the researchers and everyone who has stumbled into a swarming mound of fire ants when he said, "This is not an infestation, it's an invasion!"

The Monarch Butterfly: Jeweled Wings and Dancing Colors

Monarch butterfly
Danaus plexippus

In the wondrous world of nature, one of the most intriguing groups of animals has the flight agility of birds and the gorgeous colors of a field of wildflowers. These delicate creatures are painted with intricate patterns in a rainbow of hues. Who can doubt the creative force within the world of nature when a butterfly goes fluttering by?

The monarch reigns as the most popular and most recognized butterfly in North America. The life of this fragile animal proves that the natural world is full of hidden mystery and majesty. The first surprise is that monarchs migrate thousands of miles between their wintering grounds in the mountains of Mexico and their breeding grounds in the prairie states extending from North Texas to southern Canada. Each spring and fall, hundreds of millions of monarchs come flitting through Texas. The monarchs reach Texas by mid-March and Canada by June. They are the only butterfly with a true back and forth migration between wintering and breeding grounds. Many species disperse as their populations increase and follow the seasonal flower blooms. About 175 species migrate into Central Texas from September to November to feed on the fall wildflowers.

From late September to early November, hordes of monarchs sweep into Texas heading south for the winter. A 20-mile-wide front of the butterflies, resembling a multicolored flying carpet, flew over Dallas one day late in October. As evening fell, they flocked into trees and covered the barren limbs like leaves fluttering in a gentle breeze. As soon as the air warmed the next day, they began leaving and had departed by noon. Within 24 hours, an estimated 1,055,000 had passed over Big D. Monarchs can fly 30 mph, but between breaks for sipping nectar they typically cover only about 40 miles per day.

Whether these delicate creatures so easily swept away by wayward breezes migrated kept the scientific community puzzled for decades. Since insects can overwinter as eggs, larvae, pupae, and adults, no one was sure if the monarchs actually left or overwintered in one form or other. Studies eventually documented that they did leave their northern breeding grounds in the fall and return in the spring, but where they wintered remained a mystery. In 1952, Dr. Fred A. Urquhart of the Toronto Museum in Canada began a massive banding program. Each year, people from across the continent glued tiny stickers with the museum's address and a code number on the wings of tens of thousands of butterflies. Eagle Pass, where monarchs from across the central states funneled through Texas on their way to Mexico, proved to be a prime location for banding during the study. In good years, monarchs covered the trees like colorful leaves. In October 1957, volunteers banded almost 10,000 monarchs en route to Mexico.

As the data accumulated, the migration mystery began to become clearer. Monarchs from northern Wyoming, Montana, and Alberta—the Northern Great Basin—migrate to the Pacific Coast of California. In northern California, with cold temperatures, the butterflies congregate on tree limbs in massive numbers and hibernate. Butterflies cannot fly if the temperature drops below 55 degrees F, and they do not feed until the temperature exceeds 60 degrees. They can withstand mild frosts. Millions cover the trees in the city of Pacific Grove on the Monterey Peninsula. In warmer southern California, they do not need to hibernate and remain active all winter.

Monarchs from the eastern part of their breeding range migrate through Texas and to the Gulf Coast, where many remain active all winter. Only at Lighthouse Point, Florida, do they gather in "butterfly trees." Despite the multitudes of banded butterflies, no one knew the exact destination of the millions that flew through Texas. Finally, in 1975, the mystery of the century was solved—after 40 years of research, Urquhart discovered the winter destination in a high mountain valley in the state of Michoacán, Mexico, west of Mexico City. Hundreds of millions of monarchs, some from 2000 miles away in Canada, cover the trees. The density of the hibernating butterflies reaches four million per acre. If the winter bivouac had not been discovered, it and the monarchs that live there might have disappeared forever. The local populace depends on logging for its livelihood, and would have timbered the winter roosts if the government had not declared a logging ban. Now armed guards protect the butterfly colonies from curiosity seekers and the wrath of unemployed workers and landowners.

The beginning point and destination of the migrating monarchs tell only a fraction of the complex life cycle of these gems of nature. From late winter to early spring, monarchs begin the return trip to their summer homes. They pass through Texas from mid-March to early July. Monarchs mate soon after beginning the long trip north, and they start laying en route as soon as the eggs develop. The eggs hatch and mature, and the offspring follow their parents. Most of the monarchs that arrive in the northern breeding grounds in June and July are the generations that hatched along the migration route. They have never seen their summer destination or their Mexican wintering grounds, but they have the road map indelibly imprinted in their chromosomes.

Millions of monarchs die during the spring and fall migrations, and millions more die while hibernating. With such a high mortality, each female monarch must reproduce prodigiously if the species is to survive. Each female lays about 400 eggs, and the young mature and begin laying in about 30 days. Monarchs in the southern portion of their breeding range produce three or four generations, while northern populations produce one or two. If only half the eggs from a single female survived with half being males, the progeny of each migrating female would number two million within three generations, or about three months. Evidently the survival rate is much less than half or the world would be overrun with butterflies.

A female monarch, like most butterflies, is very particular about where she lays her eggs. Certain milkweed plants provide the perfect day care center for a baby monarch. The monarch's breeding range coincides with the North American distribution of the *Asclepias* genus of milkweed, especially *Asclepias syriaca*, and extends primarily between the latitudes 32 degrees N and 48 degrees N. Monarchs find a majority of the milkweed species in the state unsuitable for their progeny, and most pass through South and Central Texas without laying. Some do remain in the Davis Mountains, South Texas, and North Texas and spend their summer flitting from flower to flower and searching for the plants they require to lay their eggs.

The female selects a young milkweed with new leaves so the caterpillars will have tender vegetation to eat. The taste sensors on her legs tell her she has selected the right plant. She lays each egg, about the size of a pinhead, on the bottom side of a leaf. Shaped like exquisite gemstones, the eggs have about 36 facets and 22 vertical ridges. At 70 degrees F, the eggs hatch in three to four days.

The monarch passes through four distinctively different life stages: egg, caterpillar or larva, pupa, and adult. A caterpillar is designed for one purpose—eating. At hatching, the 2-mm monarch larva begins munching milkweed leaves and doesn't stop until it is about to pop, literally. The caterpillar outgrows and sheds its skin four times during its first two weeks. Then its eating machine turns off, and it takes its last bite. The caterpillar spins a tiny silk pad attached to a protected leaf or twig. As the pupa begins to form within its skin, it secures its rear feet in the pad and sheds for the last time. The new skin hardens into a protective cover called a chrysalis. Inside the chrysalis, the magical metamorphosis takes place. In about 12 days, the chrysalis bursts open and a creature as different from a caterpillar as a worm is from a bird crawls out. But the butterfly's maiden flight must wait until it pumps up its wings and the delicate membranes dry. After a life of crawling, the butterfly is earthbound no more.

A monarch's flight may seem weak and haphazard, but in reality the butterfly is a strong, purposeful flier. While searching for nectar-rich flowers, it saves energy by gliding on the fickle air currents. It cruises with a wingbeat covering 30 degrees, while its 30 mph power flight, used to escape danger and during migration, sweeps through 120 degrees. While courting, the male chases the female in an upward spiral. If the lady is unreceptive, she flees with a rapid zigzag flight. The male tries to appear appealing by wearing a sweet-smelling perfume. It even has a special pocket in its hind wings to store the scent, which is produced by glands on the tip of its abdomen.

A butterfly may seem like a tasty morsel to a hungry bird, but the monarch has a secret weapon. The caterpillar concentrates toxins in its system from the milkweed leaves it so voraciously consumes. The compounds, cardiac glycosides, are powerful heart drugs for vertebrates, and cause vomiting as well. Each monarch has three to nine times the dose of cardiac glycosides prescribed for a human with congestive heart failure. The poison is concentrated in the monarch's wings, the most likely part to be attacked. Tests with blue jays showed that a naive bird that had never seen a monarch readily attacked, but became ill after one bite. One nibble gave the bird a retching experience it never forgot. With the bird overcome with nausea, the monarch flies away only slightly scarred from the escapade. Many monarchs in the wild have nibble marks on their wings, showing they have survived a close encounter.

The viceroy butterfly, *Limenitis archippus,* which has a similar outline and the same basic coloration, benefits from the monarch's bad reputation. A bird that has tasted a monarch will avoid any butterfly that resembles the distasteful species. The ruse fools birds so well that they avoid the unprotected look-alike. Viceroys are masters of deception and use a different disguise through each of their four life stages. Their eggs resemble leaf galls, the caterpillars have fierce-looking horns, and the chrysalis looks like a bird dropping. Many species of butterflies mimic butterflies and other objects, especially in the tropics.

Monarchs are a marvel of nature. Though so lightweight and easily swept by the wind, they migrate thousands of miles. They appear defenseless, yet possess a powerful deterrent to predators. They reproduce by the millions, and when hibernating cover trees like colorful ribbons. Monarchs play an important role as pollinators as they sip the sweet nectar provided by flowers. Besides their ecological importance, butterflies have an aesthetic importance for humans. Whether we are young or old, butterflies delight our senses. Children thrill when one lands on their shoulder, and adults, with expensive cameras in hand, chase the elusive insects through fields and meadows. I'll never forget my daughter's excitement when she saw a butterfly flitting from flower to flower: "Look, there goes a flutterby!"

The Portuguese Man-of-War: Danger Rides the Surf

Portuguese man-of-war
Physalia physalis

Long before the Portuguese, or any other human, hoisted a sail, the Portuguese man-of-war ran freely before the ocean winds. Driven by breezes, currents, and tides, armadas of these boneless sea creatures landed on the sandy shores of the New World. Men-of-war sail the seas in the true sense of the word. They have a gas-filled float that serves both as a boat to keep the animals from sinking and as a miniature sail. The float, resembling a small balloon, catches the wind in a flattened scalloped ridge along the top.

Today, as in times past, Portuguese men-of-war, or blue jellyfish as they are sometimes called, ride the rolling surf, which eventually casts them up to die on dry land. Sometimes the receding tide leaves behind thousands of the unusual sea animals littering the beach. At times men-of-war seem to invade the beach, their blue-tinted floats bobbing up and down in the waves. Swimmers learn that, like a disguised pirate vessel, these innocent-looking mariners are more than meets the eye. Below the 4-to-12-inch balloonlike float dangle tentacles that present a serious threat to humans.

Portuguese men-of-war belong to a group of aquatic invertebrates called Cnidaria, which includes jellyfish, sea wasps, and other floating creatures. They have no skeleton, organs, blood, or central nervous system. One distinguishing feature makes this group important to humans: they have tentacles covered with stinging cells to kill the fish they eat. The tentacles trail through the water and automatically discharge their poison into any creature they encounter, whether fish or human. Each stinging cell, or nematocyst, less than 50 microns (0.002 inch) in size, contains a coiled, hollow thread in a tiny capsule. When mechanically disturbed, the barbed thread shoots into the victim, releasing its poison. The venom has a paralyzing effect on the motor nerve endings, which stuns or kills small fish. The tentacles contract to bring the fish up to the mouth and digestive cavity of the animal. A man-of-war's numerous tentacles contain about 900 nematocysts per inch.

Humans encountering the tentacles experience immediate and excruciating pain. Minor cases may involve intense burning, stinging, and formation of red welts. Extensive contact can result in throbbing or shooting pains, muscular cramps, nausea, vomiting, abdominal rigidity, severe backache, inability to speak, frothing at the mouth, constriction of the throat, respiratory difficulties, paralysis, delirium, convulsions, and unconsciousness. Swimmers are the most frequent victims. The stinging cells on the tentacles of men-of-war washed ashore may remain potent for several days. At least one human death, a child in Florida, has been attributed to the Portuguese man-of-war. Other fatalities have been reported, but such cases are difficult to document.

First aid for a sting from a man-of-war, jellyfish, or sea wasp can help relieve pain and partially alleviate the effects of the toxin. Splash generous amounts of rubbing alcohol on the affected area to deactivate any stinging cells that have not discharged their poison. Sprinkle the sting area with liberal amounts of unseasoned Adolph's Meat Tenderizer (or any brand containing papain) to help neutralize the toxin. The victim may require antihistamines to lessen allergic responses, and in severe cases should seek immediate medical attention. Intravenous injections and other support measures may be required to combat pain, muscle cramps, and respiratory and cardiac problems.

The tentacles of a large man-of-war may trail 100 feet beneath the animal as it drifts aimlessly across the sea, though most that wash up on the beaches have tentacles 5 to 10 feet long. The gas-filled float of this unusual dweller of the deep is a remarkable adaptation enabling a sea animal unable to swim to navigate the waters and search for food. A gas-producing gland fills the float with just the proper amount of gas to keep the man-of-war buoyant. A valve on the top regulates the pressure inside the bladder, and if it becomes deflated, the gas glands can fill the float within 30 minutes.

The Portuguese man-of-war has a complex life history, and the animal we see floating in the surf represents only half the story. Aristotle was unsure about these strange creatures and considered them an intermediate life form between plants and animals. Actually the man-of-war passes through two radically different stages of life, known to scientists as alteration of generations. The floating organism with dangling tentacles is called a medusa, after the Greek mythological woman with snakes for hair. The medusa itself is a colony composed of four distinct types of individuals, each specialized for different functions. The float keeps the colony buoyant; the feeding tentacles bring the food to the digestive chamber on the bottom side of the float; the stinging cells reside in the feeding tentacles; and the reproductive members produce the eggs and sperm.

The medusa generation reproduces sexually, but the offspring bear little resemblance to the parents. The free-swimming larvae of the second generation attach permanently to an underwater surface and grow into a polyp, a tubelike creature resembling a plant. The polyp branches and forms many slender stems. Some stems have tentacles to capture food, and others bud and break off to become the free-swimming medusae.

In the unpredictable world of the deep, some creatures have learned to use to their advantage the most effective poisons of other animals. Even the deadly tentacles of the man-of-war become a haven of safety for the small harvester, or man-of-war fish, the *Peprilus*. This fish may not be immune to the poison, but it somehow has conditioned the floating creatures not to sting it. This type of association between two widely different creatures in which one or both benefit from the characteristics of the other is widespread in the ocean. Another fish has learned to take advantage of the potent stinging cells, but at the expense of the man-of-war. It eats the tentacles without discharging the nematocysts, and somehow manages to store the armed cells in its body to use for its own defense.

Some of the most unusual inhabitants of the earth live in the sea, and many can be found along the crescent-shaped coast of Texas. At low tide, look on the exposed rocks of jetties for anemones with their waving tentacles. In addition to sand dollars and shells with intricate patterns, beachcombers find colorful jellylike creatures, sea cucumbers, and other bizarre animals that hint of a complex life structure just beyond the surf. Scientists exploring the deepest ocean trenches find animals far stranger than the most imaginative fiction. Portuguese men-of-war live in a mysterious world where nature's creative force holds unbounded surprises. No wonder Aristotle could not decide whether many of these complicated creatures belonged to the animal or plant kingdom.

The Scorpion: Miniature Monster

Striped scorpion
Centruroides vittatus

How would you like to live in a world teeming with giant flesh-eating, blood-sucking monsters? Humans are lucky. By some quirk of fate, all the monsters are too small to present much of a threat. But try looking at life from an insect's viewpoint. Miniature monsters more awesome than any found in science fiction stories populate the world around us. Behind every twig and under every rock, some vicious beast lies in ambush with claws, teeth, or stinger ready to pounce. Yet one of these minuscule monsters, despite its two-to-three-inch size, strikes fear in the heart of every human. The scorpion, with tail poised ready to strike, gets as much respect as a deadly rattlesnake.

Scorpions come well equipped to survive in the world of the small. Since their ancestors adapted to life out of the sea, about 400 million years ago according to fossils, these critters have faced few foes their own size. With powerful lobsterlike pincers and a venom that instantly kills their prey, scorpions stalk through leaves, rocks, and vegetation. When the sun sets, scorpions emerge to hunt spiders, centipedes, crickets, and other small insects. By day, the reclusive animals repose under rocks, fallen logs, and other objects. Despite their effective armament, scorpions have learned to stay hidden in a world populated with creatures that can crush them with a single step, or devour them with a snap of their jaws. Armadillos, skunks, opossums, and pallid bats consider scorpions, poison stingers and all, tasty tidbits.

While stalking prey through the darkness of night, scorpions depend not on sight but on the senses of touch and smell to find their victims. Some species also have well-developed hearing. Comblike chemical receptor organs on their undersides contact the ground as they crawl and help them track prey. Scorpions have long hairs, or setae, on their eight legs and pincers that are super-sensitive to touch. These hunters walk with claws outstretched feeling for small animals. The hairs on the legs of the sand scorpion in the Mojave Desert detect movement and direct the animal toward its meal. Sand scorpions use a complex sensing system to measure the time difference between compression waves caused by insects walking across the sand dunes. These perceptive creatures can detect movement 18 inches away, and can judge perfectly the distance and direction of prey, whether crawling or burrowing, within a 4-inch range.

A scorpion catches food by grabbing and crushing its prey with its powerful pincers. With lightning action, the tiny predator swings its tail over its body and stings the victim. The creature jerks convulsively and, paralyzed by the venom, dies in the scorpion's rigid grasp. The scorpion then chews the body into a semi-liquid state that it can suck up with its tiny mouth.

The poison glands in the swollen tip of the scorpion's tail secrete a venom that is deadly to insects and causes extreme discomfort in humans. Yet, except for a few species, none of which occur in Texas, scorpions present no serious threat to humans. Arizona harbors two species with a potent neurotoxin that can kill humans, particularly children. About two dozen deaths have been recorded. A more deadly scorpion, which reportedly can kill an adult in one hour, lives in the state of Durango in central Mexico. In a 35-year period, 1600 people died from the Durango scorpion. The government placed a bounty on the scorpions and collected 100,000 annually, but the effort did little to reduce the scorpion population. Fortunately, an antivenin that now saves many lives has been developed for the Durango and Arizona scorpions.

The sting from a nonlethal scorpion burns like fire and

leaves a welt, but like a wasp sting, the pain eventually subsides. Cooling the area with ice and taking antihistamines if an allergic reaction develops are the only recommended medical treatments. In hypersensitive individuals, however, the sting from a nonlethal scorpion like the species in Texas can cause severe complications and even death. If swelling or pain persists or breathing difficulty occurs, contact a physician immediately. A Johnson City man, stung on the finger while carrying firewood, died before he reached the hospital.

Scorpions, along with spiders and ticks, are members of the Arachnid family. Some tropical species grow to 7 inches in length, but in Texas, the common striped scorpion, *Centruroides vittatus,* only reaches 2½ inches. The 20 to 30 species in the United States live as far north as Canada, but favor the Gulf Coast and the hot, arid Southwest. Scorpions tolerate a limited range of temperatures and humidity, even in the desert. These creatures are perfect examples of animals that find a micro-habitat exactly suited to them. Like humans inside an air-conditioned building, scorpions avoid the daytime heat by staying protected under a rock or other object. The sand scorpion burrows 4 inches into the sand, where the temperature is about 60 degrees F lower than the surface and the humidity approaches 90 percent. Only in the cool of night do scorpions venture forth.

Scorpions do not socialize. Besides accidental encounters that usually result in cannibalism, only the sexual urge brings two scorpions together. They have an elaborate courtship that may last for hours: the two grasp each other's pincers and jaws and dance back and forth. Finally the male deposits a sac of sperm on the ground and pulls the female over it. She picks up the capsule with a special organ on her abdomen. After birth, the young crowd onto the mother's back and remain there until they shed their first skin. The sight of the babies covering the female's back inspired the folktale that the young eat the parent.

Several close relatives of scorpions resemble them but

do not sting. The whip scorpion has the same body shape without pincers and a long whiplike tail with no stinger. The vinegarroon also has a whiplike tail with no stinger, but has heavy pincers. The pseudoscorpion has pincers but no tail at all.

I spent one enjoyable summer swimming in Lake Travis and turning over rocks collecting scorpions for a study of their behavior. I used a black light to find the prowling predators at night. Their bodies fluoresce beautifully in the darkness, but a hand-held black light must be within several feet before the Texas species begin to glow. Scorpions are not desirable pets, since they can squeeze through the slightest crack. My entire collection disappeared from the aquarium and, fortunately, never appeared again.

To avoid scorpions around your home, move stacks of logs, rocks, and trash away from buildings and play areas. They seem particularly attracted to rocks and masonry. Carefully examine rocks and firewood before picking them up. Since scorpions take refuge under bark, delay bringing firewood into the house until you are ready to burn it.

The Tarantula: The Mild-mannered Mygalomorph

Red-legged tarantula
Dugesiella hentzi

Suddenly a shriek of alarm rang through the woods, and I immediately thought "RATTLESNAKE." The group on my nature hike had parted like the Red Sea, as one of the most feared critters on earth crawled across the path. Someone yelled, "Get a stick, kill it!" I decided to take a chance and reached down and scooped up the fearful-looking creature. I heard gasps of disbelief as the animal slowly crawled across my palm and over my wrist.

"Does anyone want to hold a tarantula?" I asked casually. Finally, one brave lady held out her hand and let the hairy spider tiptoe across her arm. The demonstration helped dispel another common misconception about the animal. Despite their reputation, tarantulas are not deadly, although their large fangs can deliver a painful bite. Even though poisonous to insects, the venom of these giant spiders of the suborder Mygalomorphae presents no more of a threat to humans than a mosquito bite, or a bee sting at worst.

Ironically the tarantula's true nature bears little resemblance to its reputation. These secretive, ground-dwelling spiders spend most of their life hidden in their burrows. Unlike some spiders, tarantulas have poor vision and, even with eight eyes, cannot see farther than a few inches. They cannot run fast or jump more than the length of their body. Tarantulas, aided by a thick covering of sensitive hairs, communicate with the world mainly through the sense of touch.

The pet industry long ago discovered the gentle nature of tarantulas, and pet stores often sell them as novelties. They don't bark, have smelly litter boxes, or require property deposits. But they may cause heart attacks and strain friendships. My tarantula, Legs, had been missing for three days when a friend came to visit. She left in a huff when, like little

Miss Muffet, she found Legs sitting beside her.

When alarmed, tarantulas can appear as threatening as their evil reputation suggests. They rear back, wave their long, hairy legs, and expose the formidable fangs on their underside. These frightful-looking creatures have a secret weapon to complement their fearsome posturing and powerful fangs: a distinctive patch of poisonous hairs on their abdomen. If attacked, they rapidly scrape off a cloud of the tiny hairs that irritate the eyes and nose of their antagonist. The hairs penetrate the mucous membranes of mammals, and can cause a rash, or even partial blindness. Before the attacker recovers, the spider dashes to the safety of its burrow. Some Mexican species can cause swelling and stinging in humans.

Besides being the largest spiders, tarantulas live longer than any other terrestrial invertebrates. They take 8 to 10 years to mature sexually. The females may live for 30 years, but the males, who do not molt after becoming an adult, die within a year after they mature. Southwestern tarantulas spend most of their long lives in or within a few meters of their burrows, which they may call home for several decades. At the slightest sign of danger, they hide in their hole and cover the entrance with a silken tapestry. The mature males gad about during the mating season. Virtually all the tarantulas seen crossing roads between June and December are males searching for females.

Skunks and armadillos, unaffected by the fear complex so prevalent in humans, view tarantulas as culinary delicacies. They root them out of the ground, along with scorpions and other equally tasty critters. Birds also dine on the spiders, especially the smaller juveniles. In addition to hungry mammals and birds, tarantulas have some unconventional enemies that do not fight by the rules of the Geneva Convention. Two of the tarantula's most insidious enemies want the large spider alive, not dead. To them, the eight-legged creature represents a living food larder, a supermarket for their offspring. The tarantula's large size and powerful fangs do not always ensure victory against smaller foes. Tarantulas must contend with adversaries that use time bombs and a poison that places them in suspended animation. Small-headed flies of the family Acroceridae lay their eggs on the bodies of tarantulas. Like tiny time bombs, the eggs eventually hatch and the maggots voraciously devour the hapless spider.

The tarantula hawk, a wasp in the genus *Pepsis,* hunts the large spiders like a cat after a mouse. When the wasp encounters a tarantula, the two spar like a pair of gladiators. The wasp circles and darts in trying to avoid the deadly fangs of her larger opponent. In this fight to the death, the tarantula has no retreat. The wasp will pursue the spider right into its burrow. Eventually the wasp either ends up in the grips of her intended victim or delivers a paralyzing sting to the spider. Once stung, the tarantula does not die, but may live for several months unable to move. The victor drags the bulky tarantula to a burrow she has excavated and deposits it with her eggs. When the eggs hatch, they have a ready supply of fresh food.

Usually the tarantula plays the role of the terrorist. In the summer, the hungry predator ventures out every night in search of food. Beetles, grasshoppers, sow bugs, millipedes, and other spiders and insects that wander into the giant's neighborhood seldom survive. The tarantula, like all spiders, predigests its victim by flooding the dead creature with enzymes. The spider then sucks up the softened tissue. South American species, larger than our tarantulas, devour frogs, toads, lizards, and even mice. A captive South American speciman fed for 24 hours on a mouse, leaving only a shapeless, empty skin.

The largest tarantulas live in the tropics. South Americans call them bird spiders because the giant creatures sometimes catch birds. One even killed a small rattlesnake by pouncing on its head. A Brazilian species has a 3½-inch body, a 9½-inch leg span, and weighs 3 ounces. Another species in Guiana has a 10-inch leg span. About 30 species of tarantulas live in the United States, mostly in the arid Southwest. The largest has a 2-inch body, 6-to-7-inch leg span, and weighs half an ounce.

No description of this hairy, long-legged spider that looks like some creature from a horror movie would be complete without mentioning its bizarre sex life. Spiders in general have strange sexual practices, especially the species with females that may eat the smaller males during the act of mating. Both sexes of the tarantulas are about the same size, but the males still must be cautious of the ladies' temperament and appetite. Females often kill males during courtship and mating.

The transfer of sperm from the male to the female requires specialized equipment for both sexes. The male has two specialized claws shaped like syringes on the ends of its two pedipalps, appendages on its head. He weaves a little silken purse to hold a globule of sperm, and places a package in each syringe on his pedipalps. The female has two pouches on her abdomen designed to receive the sperm packages. She can store the sperm for weeks or even months. As she lays her eggs, she bathes each one in a fluid containing the sperm.

North American tarantulas lay about 600 to 1000 eggs at a time on a silken sheet that the female weaves. The spider covers the eggs with another sheet and seals the edges. The large, flabby bags are about three inches in diameter. The female takes the bag to the surface of the burrow to warm in the sun, and carefully guards her treasure for about seven weeks until hatching. The babies scatter to dig their own burrows and battle the odds of survival. The vulnerable young suffer greatly from predation, and few reach sexual maturity. When they finally mature, the males have longer legs than the females, are darker, and have reddish hair on their abdomens. Some of the South American giant species mature in three to four years.

The mygalomorph spiders received their common name from a large wolf spider named after the town of Taranto in southern Italy. European explorers thought the giant spiders of the New World resembled the tarantulas in Italy, and gave them the same name. During the Middle Ages, epidemics of tarantism, or spider bites, swept through southern Europe. Music provided the only antidote. The victims gathered in an elaborately decorated hall to perform the tarantula dance, the tarantella, to the cheers and encouragement of a sympathetic audience. For an effective cure, the participants had to leap, jerk, and dance incessantly until perspiration drove out the poison. The cure was complete when the dancer dropped from exhaustion. For a fee, villagers often allowed spiders to bite them, so tourists could witness the amazing cure. The event had striking similarities to pagan exhibitions prohibited by the Catholic Church.

People almost universally detest spiders, and particularly tarantulas. Poems and nursery rhymes terrify children. Hollywood even showed macho super spy James Bond paralyzed with horror when a hairy tarantula crawled across his chest. Except for the black widow and brown recluse, which have particularly toxic venom, spiders rarely intrude into human life beyond cobwebs in the attic. But even though the tarantula has a gentle disposition, you had better avoid this giant spider. According to Zuni legend, the tarantula is an expert trickster and will trick you out of your clothes if you are not careful.

Texas' Vanishing Wildlife: Endangered and Threatened Animals

MAMMALS

Extinct Species

Black-footed ferret, *Mustela nigripes:* poisoning of prairie dogs, its main prey; presumed extinct until 1981, when several were discovered in Wyoming; only 25 remaining in captivity.

Gray wolf, *Canis lupus:* habitat destruction, eradication of bison, predator control.

Texas grizzly bear, *Ursus texensis:* eradicated in 1890.

Merriam elk, *Cervus elaphus merriami:* extinct subspecies.

Texas mountain sheep, *Ovis canadensis texiana:* extinct since the 1950s; different subspecies now introduced in Trans-Pecos.

Endangered Species

Jaguar, *Felis onca:* habitat destruction, hunting, predator control.

Jaguarundi, *Felis yagouaroundi*: habitat destruction; not sighted in Texas in ten years.

Manatee, *Trichechus manatus:* occurs in Florida and Caribbean, historically has ventured as far west as Texas.

Ocelot, *Felis pardalis:* habitat destruction, hunting, road kills.

Red wolf, *Canis rufus:* habitat destruction, predator control, hybridization with dogs; very few left in the wild.

Black bear, *Ursus americanus:* incompatible with humans; few or none remaining in mountains of Trans-Pecos.

Coati, *Nasua nausa*: very few occur in Texas since the southern portion of the state is the northern limit of its range.

Threatened Species

Rafinesque's big-eared bat, *Plecotus rafinesquii:* habitat loss, pesticide use.

Spotted bat, *Euderma maculatum:* habitat destruction.

Southern yellow bat, *Lasiurus ega:* habitat destruction.

Atlantic spotted dolphin, *Stenella plagiodon:* commercial fishing.

Palo Duro mouse, *Peromyscus comanche:* habitat loss to wildfire and impounded lakes.

Texas kangaroo rat, *Dipodomys elator:* habitat loss to agriculture and urbanization.

Coues' rice rat, *Oryzomys couesi:* habitat loss to agriculture.

BIRDS

Endangered Species

Whooping crane, *Grus americana:* habitat loss.

Eskimo curlew, *Numenius borealis:* migrates through Texas.

Bald eagle, *Haliaeetus leucocephalus:* pesticide use, hunting; about 20 breeding pairs occur in Texas.

American peregrine falcon, *Falco peregrinis anatum:* pesticide use.

Brown pelican, *Pelecanus occidentalis:* pesticide use.

Greater Attwater's prairie chicken, *Tympanuchus cupido attwateri:* habitat loss.

Interior least tern, *Sterna antillarum athalassos:* nesting disturbance by human recreation, water control projects.

Ivory-billed woodpecker, *Campephilus principalis:* habitat loss, probably extinct.

Red-cockaded woodpecker, *Picoides borealis:* habitat loss from lumbering.

Aplomado falcon, *Falco femoralis:* no recent breeding in Texas.

Black-capped vireo, *Vireo atricapillus:* habitat loss to urban development, nest parasitism by brown-headed cowbirds.

Threatened Species

Arctic peregrine falcon, *Falco peregrinus tundrius:* breeds in Canada but winters in Texas.

Reddish egret, *Egretta rufescens:* human disturbance of nesting sites.

Common black hawk, *Buteogallus anthracinus:* habitat loss.

Gray hawk, *Buteo nitidus:* habitat loss to agriculture.

White-tailed hawk, *Buteo albicaudatus:* loss of nesting habitat.

Zone-tailed hawk, *Buteo albonotatus:* habitat loss.

White-faced ibis, *Plegadis chihi:* pesticide use.

Swallow-tailed kite, *Elanoides forficatus:* a Texas migrant.

Ferruginous pygmy owl, *Glaucidium brasilianum:* breeding habitat loss to agriculture.

Wood stork, *Mycteria americana:* no longer breeding in Texas.

Golden-cheeked warbler, *Dendroica chrysoparia:* habitat loss, nest parasitism to brown-headed cowbirds.

Rose-throated becard, *Pachyramphus aglaiae:* habitat loss to agriculture.

Tropical parula, *Parula pitiayumi:* habitat loss to agriculture.

Bachman's sparrow, *Aimophila aestivalis:* habitat loss to lumbering.

Botteri's sparrow, *Aimophila botterii:* habitat loss.

Sooty tern, *Sterna fuscata:* nest disturbance by human recreation.

Northern beardless tyrannulet, *Camptostoma imberbe:* habitat loss.

REPTILES

Endangered Species

Atlantic hawksbill turtle, *Eretmochelys imbricata imbricata:* hunted for shell, drowns in fishnets.

Leatherback turtle, *Dermochelys coriacea:* occurs infrequently in Texas waters, drowns in fishnets.

Speckled racer, *Drymobius margaritiferus margaritiferus* (snake): common in the tropics but barely persisting in South Texas because of habitat loss.

Kemp's ridley turtle, *Lepidochelys kempi:* egg vandals at the Mexican nesting site, drowns in fishnets; attempts to establish a nesting colony on Padre Island National Seashore still inconclusive.

Concho water snake, *Nerodia harteri paucimaculata:* habitat loss to impounded lakes.

Loggerhead turtle, *Caretta caretta:* once abundant, now seen infrequently; drowns in fishnets.

Northern cat-eyed snake, *Leptodeira septentrionalis septentrionalis:* habitat loss, pesticide use.

Louisiana pine snake, *Pituophis melanoleucus ruthveni:* habitat destruction by lumbering.

Big Bend mud turtle, *Kinosternon hirtipes murrayi:* water depletion, habitat loss.

Western smooth green snake, *Opheodrys vernalis blanchardi:* habitat loss, pesticide use.

Threatened Species

Brazos water snake, *Nerodia harteri harteri:* dam construction.

Reticulated gecko, *Coleonyx reticulatus (lizard):* pesticide use, habitat loss.

Reticulate collared lizard, *Crotaphytus reticulatus:* habitat loss, pesticide use.

Texas horned lizard, *Phrynosoma cornutum:* overcollecting, pesticide use.

Mountain short-horned lizard, *Phrynosoma douglassii hernandesi:* habitat loss, pesticide use.

Black-striped snake, *Coniophanes imperialis imperialis:* habitat loss to agriculture.

Texas indigo snake, *Drymarchon corais erebennus:* habitat loss to agriculture, overcollecting.

Texas lyre snake, *Trimorphodon biscutatus vilkinsonii:* overcollecting, habitat loss.

Baird's rat snake, *Elaphe bairdi:* over collecting.

Texas Tortoise, *Gopherus berlandieri:* overcollecting, habitat loss.

Atlantic green turtle, *Chelonia mydas mydas:* over collecting, drowns in fishnets.

Big Bend blackhead snake, *Tantilla rubra:* habitat loss, pesticide use.

Texas scarlet snake, *Cemophora coccinea lineri:* overcollecting, habitat loss.

Northern scarlet snake, *Cemophora coccinea copei:* habitat loss, overcollecting.

Alligator snapping turtle, *Macroclemys temminckii:* habitat loss, commercial exploitations, pesticide use.

Books for Further Reading

Bedichek, Roy. *Adventures with a Texas Naturalist*. Austin: University of Texas Press, 1967.

Bent, A. C. *Life Histories of North American Birds of Prey*. New York: Dover Publications, 1961.

———. *Life Histories of North American Flycatchers, Larks, Swallows, and Their Allies*. New York: Dover Publications, 1961.

———. *Life Histories of North American Marsh Birds*. New York: Dover Publications, 1961.

———. *Life Histories of North American Nuthatches, Wrens, Thrashers, and Their Allies*. New York: Dover Publications, 1961.

Breland, Osmond P. *Animal Life and Lore*. New York: Harper and Row Publishers, 1972.

Brown, Leslie. *Eagles*. New York: Arco Publishing Co., 1970.

Burt, William H. *A Field Guide to the Mammals*. New York: Houghton Mifflin Co., 1964.

Burton, John A. *Owls of the World*. New York: A & W Visual Library, 1973.

Carrington, Richard. *The Mammals*. New York: Time-Life Books, 1963.

Chapman, Joseph A., and George A. Feldhammer. *Wild Mammals of North America*. Baltimore: Johns Hopkins University Press, 1982.

Cruickshank, Allan, and Helen Cruickshank. *1001 Questions Answered About Birds*. New York: Dover Publications, 1958.

Dalrymple, Byron W. *North American Game Animals*. New York: Outdoor Life, 1978.

Davis, William B. *The Mammals of Texas*. Austin: Texas Parks and Wildlife Department, Bulletin 41, 1974.

Dobie, J. Frank. *The Longhorns*. Austin: University of Texas Press, 1980.

———. *Wild and Wily Range Animals*. Flagstaff: Northland Press, 1980.

———. *A Vaquero of the Brush Country*. Austin: University of Texas Press, 1981.

———. *Rattlesnakes*. Austin: University of Texas Press, 1982.

Duplaix, Nicole, and Noel Simon. *World of Mammals*. New York: Crown Publishers, 1976.

Engelmann, Wolf-Eberhard, and Fritz J. Obst. *Snakes, Biology, Behavior, and Relation to Man*. Los Angeles: Exter Books, 1981.

Garrett, Judith M., and David G. Barker. *A Field Guide to Reptiles and Amphibians of Texas*. Austin: Texas Monthly Press, 1987.

Gertsch, Willis J. *American Spiders*. New York: Van Nostrand Reinhold Co., 1979.

Gibbons, Whit. *Their Blood Runs Cold*. University: University of Alabama Press, 1983.

Lane, Gary. *Life of the Past*. Columbus: Charles E. Merrill Publishing Co., 1978.

Layton, R. B. *The Purple Martin*. Jackson, Miss.: Nature Book Publishers, 1969.

Madison, Virginia. *The Big Bend Country of Texas*. Stonington, Conn.: October House, 1968.

Oberholser, Harry C. *The Bird Life of Texas*. Austin: University of Texas Press, 1974.

Pearson, Erwin W., and Milton Caroline. Predator Control in Relation to Livestock in Central Texas. *Journal of Range Management* 34 (November 1981): pp 435–41.

Perrins, Christopher, and C. J. O. Harrison. *Birds, Their Life, Their Ways, Their World*. New York: Reader's Digest Association, 1979.

Pulich, Warren M. *The Golden-Cheeked Warbler*. Austin: Texas Parks and Wildlife Department, 1976.

Robbins, Chandler S., Bertel Bruun, and Herbert S. Zim. *A Field Guide to the Birds of North America*, New York: Golden Press, 1983.

Rood, Ronald. *Animals Nobody Loves*. New York: Stephen Greene Press, 1971.

Schmidly, David J. *The Mammals of Trans-Pecos Texas*. College Station: Texas A&M Press, 1977.

Skutch, Alexander F. *The Life of the Hummingbird*. New York: Crown Publishers, 1973.

———. *Parent Birds and Their Young*. Austin: University of Texas Press, 1976.

Tennant, Alan. *The Snakes of Texas*. Austin: Texas Monthly Press, 1984.

———. *A Field Guide to the Snakes of Texas*. Austin: Texas Monthly Press, 1985.

Texas Animal Damage Control. U.S. Fish and Wildlife Service, Annual Report FY 1983.

Wauer, Roland H. *Naturalist's Big Bend*. Santa Fe: Peregrine Productions, 1973.

Wetmore, Alexander. *Water, Prey, and Game Birds of North America*. Washington, D.C.: National Geographic Society, 1965.

Whitfield, Philip. *The Hunters*. New York: Simon and Schuster, 1978.